Forensic DNA Collection at Death Scenes
A Pictorial Guide

Rhonda Williams, PhD
Roger Kahn, PhD

CRC Press
Taylor & Francis Group
Boca Raton London New York

CRC Press is an imprint of the
Taylor & Francis Group, an **informa** business

CRC Press
Taylor & Francis Group
6000 Broken Sound Parkway NW, Suite 300
Boca Raton, FL 33487-2742

© 2014 by Taylor & Francis Group, LLC
CRC Press is an imprint of Taylor & Francis Group, an Informa business

No claim to original U.S. Government works

Printed on acid-free paper
Version Date: 20140131

International Standard Book Number-13: 978-1-4822-0369-1 (Paperback)

This book contains information obtained from authentic and highly regarded sources. Reasonable efforts have been made to publish reliable data and information, but the author and publisher cannot assume responsibility for the validity of all materials or the consequences of their use. The authors and publishers have attempted to trace the copyright holders of all material reproduced in this publication and apologize to copyright holders if permission to publish in this form has not been obtained. If any copyright material has not been acknowledged please write and let us know so we may rectify in any future reprint.

Except as permitted under U.S. Copyright Law, no part of this book may be reprinted, reproduced, transmitted, or utilized in any form by any electronic, mechanical, or other means, now known or hereafter invented, including photocopying, microfilming, and recording, or in any information storage or retrieval system, without written permission from the publishers.

For permission to photocopy or use material electronically from this work, please access www.copyright.com (http://www.copyright.com/) or contact the Copyright Clearance Center, Inc. (CCC), 222 Rosewood Drive, Danvers, MA 01923, 978-750-8400. CCC is a not-for-profit organization that provides licenses and registration for a variety of users. For organizations that have been granted a photocopy license by the CCC, a separate system of payment has been arranged.

Trademark Notice: Product or corporate names may be trademarks or registered trademarks, and are used only for identification and explanation without intent to infringe.

Library of Congress Cataloging-in-Publication Data

Williams, Rhonda (Rhonda Clark), author.
 Forensic DNA collection at death scenes : a pictorial guide / Rhonda Williams, Roger Kahn.
 p. ; cm.
 Includes bibliographical references and index.
 ISBN 978-1-4822-0369-1 (hardcover : alk. paper)
 I. Kahn, Roger (Forensic scientist), author. II. Title.
 [DNLM: 1. Forensic Medicine--methods--Atlases. 2. DNA Fingerprinting--methods--Atlases. W 617]

 RA1057.55
 614'.1--dc23   2014001310

Visit the Taylor & Francis Web site at
http://www.taylorandfrancis.com

and the CRC Press Web site at
http://www.crcpress.com

Contents

Foreword	v
Preface	vii
Acknowledgments	ix
About the Authors	xi

1 Introduction — 1
- Introduction and History — 1
- The Program Matures — 2

2 Training — 5
- Scene Interactions — 6
- Supplies — 7
- Documentation — 9
- Collection Techniques — 10

3 Preservation and Packaging — 37

4 Body Swabbings — 53

5 Blood Patterns — 73

6 Bindings, Clothing, Wrappings — 85

7 Tough Places — 119

8 Trace Success Cases — 127

9 Administration — 157
- Closing Remarks — 158
- References — 158

Index — 159

Foreword

Early on in my role as chief medical examiner at the Harris County Institute for Forensic Sciences (HCIFS), I recognized the importance of the collection of trace evidence from a decedent at the scene of death. The significance of this practice has escalated in recent years in step with advances in DNA and other forensic sciences. Biological and trace evidence that is properly documented and safeguarded by medical examiners, forensic investigators, scientists, and healthcare professionals is vital for the proper functioning of the criminal justice system. Evidence collected from a victim or crime scene is often essential to ensure, in equal measure, that the guilty are conclusively identified and convicted and that the innocent are protected from unwarranted prosecution.

Forensic DNA Collection at Death Scenes: A Pictorial Guide is the first atlas of its kind—a handbook devoted to trace evidence and DNA collection at the death scene rather than in the autopsy suite. Our hope is that this atlas will help minimize the considerable risk of altering or losing trace evidence during transportation of the body. The procedures outlined herein provide a practical approach with emphasis on the identification, documentation, and preservation of evidence. A wide range of cases investigated at the HCIFS, and the lessons learned by our Trace DNA Evidence Collection Team (TECT) over the past ten years, are laid out and examined to provide a road map to best practices in the field.

We recognize that the variable resources and organizational infrastructures among federal, state, and local agencies often do not permit the operation of an independent, multidisciplinary forensic facility such as HCIFS. Therefore, the full TECT approach applied in this office (utilizing scientists from the Forensic Genetics Laboratory) is impractical for some jurisdictions. Nonetheless, many of the concepts and procedures presented in this guide are designed for implementation by any medicolegal death investigation system.

While the scientific methodology in this field has become state of the art, this atlas makes clear that best practices may still be achieved with basic "low tech" collection and preservation procedures, matched with resourceful cognitive output—a combination of knowledge, perception, and judgment.

This atlas will be of benefit not only to medical examiners, healthcare professionals, forensic investigators, serologists and DNA analysts, but also to crime scene technicians, law enforcement, prosecutors, defense attorneys, judges, and forensic educators, as well as all others who are committed to sound scientific evidence and impartiality in the pursuit of criminal justice.

There is no doubt that the increasing applications of evidence-based medicine and forensic science to criminal justice and civil litigation demand that crime scene investigations be more scientific, better organized, and multidisciplinary. This atlas responds by providing a step-by-step guide to effective, uncompromising evidence collection. The development of the Trace DNA Evidence Team as an integral part of the medicolegal investigation is indeed a specialty whose time has come!

Luis A. Sanchez, MD
Executive Director & Chief Medical Examiner
Harris County Institute of Forensic Sciences
Houston, Texas

Preface

> Wherever he steps, whatever he touches, whatever he leaves, even unconsciously, will serve as a silent witness against him. Not even his fingerprints or his footprints, but his hair, the fibres from his clothes, the glass he breaks, the tool marks he leaves, the paint he scratches, the blood or semen he deposits or collects—all of these bear mute witness against him. This is evidence that does not forget. It is not confused by the excitement of the moment. It is not absent because human witnesses are. It is factual evidence. Physical evidence cannot be wrong; it cannot perjure itself; it cannot be wholly absent—only its interpretation can err. Only human failure to find it, study and understand it can diminish its value.
>
> —P.L. Kirk (1953)

Looking at the same thing from differing points of view can be interesting. A forensic pathologist examining bruises on the neck of a murder victim might consider whether the pattern is consistent with strangulation. A forensic DNA analyst looking at the same bruises might see an opportunity to find DNA linking a known suspect or identifying a new suspect with a DNA database. A multidisciplinary case approach is common in crime labs but not always so in medicolegal death investigations. In a crime lab, a latent print examiner, a trace evidence analyst, and a forensic DNA expert will meet beforehand when evidence needs to be tested in multiple ways. This ensures the work of one does not impede the work of another. A bloody fingerprint must be a fingerprint first and a DNA sample second.

Forensic pathologists, focusing on the cause and manner of death in the morgue, might miss the chance to identify an assailant from DNA left on the decedent's skin. At the crime scene, no one is likely to look for an assailant's DNA on the decedent. Death investigators focus on documenting the scene and preparing to transport the body to the morgue. Crime scene personnel are usually kept from the decedent and only collect evidence away from the body.

It is easy to miss evidence that cannot be seen. DNA left on a body, clothing, or bindings might not be as apparent as a blood or semen stain. Even close inspection of the decedent might not reveal its presence. Most of the DNA links described in the pages that follow are from touch DNA, small amounts of DNA transferred from the assailant to the decedent by touch. It is collected from the bodies using routine collection methods and it is tested with the same DNA test that most forensic DNA labs use. We have been surprised that samples collected from decedents at crime scenes frequently contain foreign DNA and in many cases the DNA links an assailant to the crime. We are pleased to share the lessons we have learned collecting DNA from bodies at hundreds of crime scenes.

Note: The photos in this book are from actual crime scenes; they may be offensive to some viewers and should be viewed with caution. Because the photos are from crime scenes, photo quality varies.

Acknowledgments

This book is dedicated to the families of the victims of homicides in Harris County (Houston), Texas. We would like to recognize the Trace DNA Evidence Collection Team analysts, past and present, for their dedication and hard work. We also thank HCIFS Deputy Chief Medical Examiner Dwayne A. Wolf, MD, PhD, for his clear vision, his helpful guidance, and his vigorous support of the team. We also thank him for invaluable comments and suggestions for this book. We also thank Kathy Haden-Pinneri, MD, and Pramod Gumpeni, MD, directors of the HCIFS Investigation Division, for their leadership in integrating the trace team into the death investigation process. Both are awakened frequently during the night to assess and approve trace team callouts. Their seemingly limitless energy is greatly appreciated. And last, we thank the Harris County Institute of Forensic Sciences Executive Director and Chief Medical Examiner Luis A. Sanchez, MD, for inspiring and supporting the Trace DNA Evidence Collection Team to ever-growing success.

About the Authors

Rhonda C. Williams, PhD, earned her doctorate in biochemistry/molecular biology from the University of Oklahoma Health Sciences Center in 2006. Dr. Williams joined the Harris County Institute of Forensic Sciences Forensic Genetics Laboratory in May 2006 as a DNA analyst. She was appointed to the DNA Trace Evidence Collection Team in February 2007 and currently serves as the team's lead. In addition, Dr. Williams serves on the Mass Fatality Committee for the Institute. Dr. Williams is a member of the Association of Forensic DNA Analysts and Administrators, the American Society of Biochemistry and Molecular Biology, and is certified as a Molecular Biology Fellow by the American Board of Criminalistics.

Dr. Roger Kahn holds a PhD in human genetics from Yale University in New Haven, Connecticut. He joined the Miami-Dade Police Department in the late 1980s to establish one of the first forensic DNA laboratories in the United States. He later served the Ohio Bureau of Criminal Identification and Investigation for nearly a decade as director of the state's three crime laboratories. Currently, he is crime laboratory director of the Harris County Institute of Forensic Sciences in Houston, Texas. Dr. Kahn is a past president of the American Society of Crime Laboratory Directors and is certified as a Fellow in Molecular Biology by the American Board of Criminalistics.

Introduction

Introduction and History

The Harris County (Houston, Texas) Institute of Forensic Sciences began to send, on occasion, a trace evidence analyst to homicide scenes in the early 2000s. Crime lab analysts from the trace evidence section collected trace tape lifts from the decedent in search of foreign hairs, fibers, and other trace evidence that might link a perpetrator to the crime. It soon became apparent that traditional trace evidence rarely contributed to investigations. Even when a suspect was known, trace evidence seldom linked the individual to the crime. In the best of circumstances, comparisons led to matching class characteristics. Occasional associations, but not identifications, resulted. What is more, investigators only occasionally submitted the trace tape lifts for analysis. Lab officials began to question the value of the trace evidence collection project and whether the benefit was worth the cost. Over time the focus changed to DNA collection, primarily touch DNA. There were several reasons for this. On a couple of occasions, a pathologist's swab of a bruise or a very light stain on a homicide victim's skin revealed foreign DNA that helped link a suspect to the crime. These cases alerted the medical examiner staff to the potential power of touch DNA.

Around the same time, in 2007, the Harris County Forensic Genetics Laboratory began to encourage the submission of evidence from property crimes. Most crime laboratories will test DNA evidence from property crimes if success is nearly certain. Bloodstains, cigarette butts, and ski masks, for example, almost always yield complete DNA results, and most labs will test these items whether they are from crimes against persons or from property crimes. Complete, or nearly complete, DNA results are needed for entry into the FBI's national DNA database, known as CODIS. The database compares the DNA of millions of previously convicted offenders to the DNA from crime scenes. A match can strongly link a previously unidentified offender to a crime.

Few labs will test property crime evidence that was merely touched by the perpetrator as the odds of obtaining a useful DNA profile are much lower. Nonetheless, in 2007 the HCIFS Forensic Genetics Laboratory began in earnest to test large numbers of touch DNA samples from property crimes. To our surprise, touched objects often provided full or nearly full DNA profiles that matched an offender in CODIS. Investigators responded by submitting increasing numbers of touch DNA property crime DNA cases and since that time nearly half of HCIFS DNA cases from property crimes are solely touch DNA evidence. As a result, the HCIFS lab leads all Texas crime laboratories in the total number of CODIS offender matches and total number of matches from property crimes. While this was taking place, the trace evidence collection team was in transition and eventually the entire team as well as the team lead were forensic genetics analysts who had firsthand experience with touch DNA. They began to swab bindings and bruises on decedents in addition to collecting trace tape lifts and they began to see surprising numbers of successes from the foreign DNA they recovered.

The Program Matures

Today, the Trace DNA Evidence Collection Team, the TECT, is an accredited component of the HCIFS Crime Lab. In 2013, ASCLD/LAB accredited the TECT in the crime scene discipline of specifically for Trace/DNA Evidence Collection. The TECT has grown to ten forensic DNA analysts who view scenes with a keen appreciation for touch DNA. These analysts focus on collecting samples from areas where the assailant may have touched the decedent during a struggle or while moving the body. They also comprehensively sample bindings and wrappings in a search for foreign DNA.

The DNA analysts of the TECT are volunteers who work week-long, 24-hour call rotations in addition to their duties in the DNA lab. The call-out criteria is formalized in order to reduce the number of unnecessary scene responses. TECT staff members are called only to suspected homicides that meet at least one of these criteria [Figure 1.1].

1. The decedent has been found in a place other than the original crime scene; i.e., the body was transferred from one location to another.
2. The decedent was found bound or wrapped by, e.g., with duct tape, handcuffs, zip ties, belts, or a comforter.
3. The decedent was killed by means that required sustained close contact, e.g., sharp force trauma, strangulation, or blunt force trauma; or the investigation suggests that such close contact occurred prior to or contemporaneous with the death regardless of the cause of death.

When a crime scene is identified for potential TECT activity, a team member accompanies an HCIFS investigator day or night, in any weather.

TECT scenes are a small subset of the deaths to which the HCIFS responds. Harris County, including the City of Houston, has a population of 4.1 million people according to the 2012 U.S. Census Bureau, making it the third most populous in the United States while Houston is the fifth most populous metro area in the United States. The county spans more than 1700 square miles. The HCIFS includes the medical examiner service and a full-service crime laboratory. The medical examiner is responsible for responding to and investigating unexplained deaths throughout the county. Deaths that must be reported to the HCIFS are specified by state law (CCP, Article 49.25) and include deaths that are suspected to have resulted from physical or chemical injury, sudden and unexpected deaths, deaths under unknown circumstances, suspected suicides and homicides, and deaths of children.

Introduction

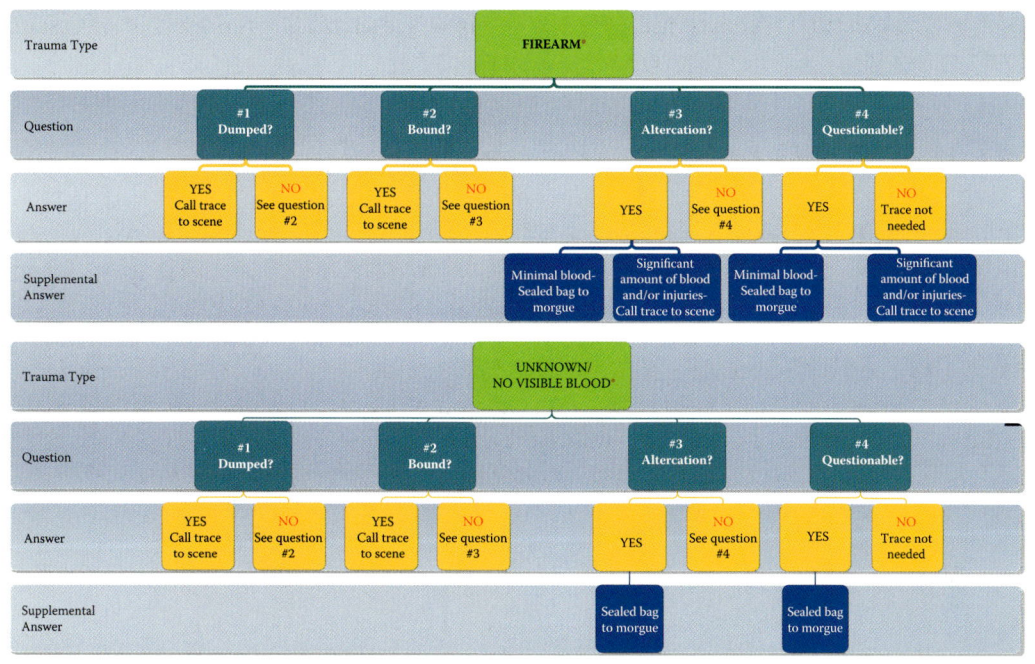

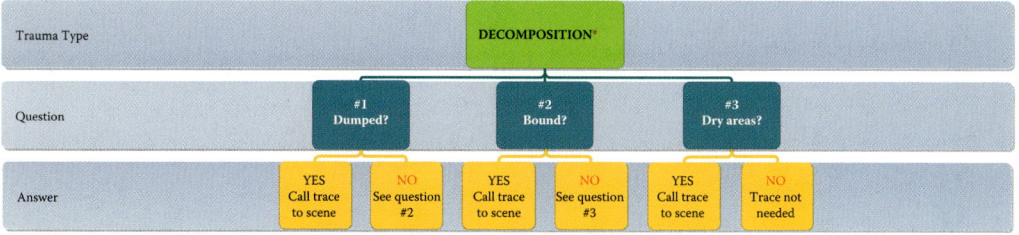

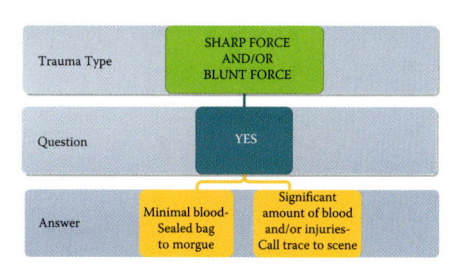

Figure 1.1 The flowchart used by investigations personnel to determine if trace collections should be performed. Depending on the questions and responses from on-scene law enforcement, the flowchart will direct the investigator toward calling the trace analyst for a collection at the scene or in the morgue.

Training

2

According to Locard's principle of exchange, whenever two objects come into contact, materials are exchanged between them (Locard, 1920). Team members are trained to search for and collect material transferred to the decedent, to the decedent's clothing, or to bindings on the decedent. To be eligible for training, a volunteer must be fully trained as a forensic DNA analyst. Once selected for training, volunteers participate in an in-depth training beginning with literature readings on trace collection theory and practice and a thorough review of TECT standard operating procedures. Hands-on training starts with a mock collection instructed by the team lead. Once the mock collection has been performed and all readings are completed, the trainee must pass two quizzes. Photos illustrating a variety of scenarios from prior scenes are reviewed with the trainee during this time. These are the photos presented in this atlas.

Next, the trainee begins to attend actual crime scenes under the supervision of a qualified TECT analyst (Figures 2.1 and 2.2). After observing two to three scenes, the trainee is permitted to process two to three scenes under supervision (Figures 2.3 and 2.4). After passing an oral exam administered by the team lead, the trainee is approved to begin independent trace collections at crime scenes.

All qualified TECT analysts participate in a proficiency testing cycle; TECT-specific proficiency tests are administered once per calendar year. The test is prepared in-house by the Harris County Institute of Forensic Sciences Quality Management section using a latex mannequin to simulate a variety of crime scene scenarios (Figure 2.5). Each TECT analyst is given one of a variety of scenarios and asked to perform a trace collection as he or she would at an actual crime scene. Scenarios include bound, stabbed, and strangled decedents. In order to complete the test successfully, the analyst must be properly gowned at the scene, complete documentation properly, use correct collection techniques for the scenario, and retrieve key items of evidence. The test is observed by the trace team lead who evaluates the work and provides a written critique to the analyst. The TECT analyst must successfully complete the test to continue performing trace collections. An unsuccessful test would result in retraining and a requalifying exam using a different case scenario.

For continuing education, the team meets quarterly to discuss case information and to gather ideas and approaches that contribute to best practices. Current literature is reviewed and discussions are held on topics relevant to trace collection practices, such as bloodstain pattern analysis and crime scene collection techniques. We review situations and consider lessons learned. Most of all, we strengthen the camaraderie that makes this team work so well.

Additionally, representatives from the team meet regularly with an advisory "board" comprised of HCIFS forensic pathologists, the crime laboratory director, and the directors of the HCIFS Investigations Division. Specific scenes are reviewed and results are presented, with an eye toward identifying which types of scenes are the most fruitful for obtaining foreign DNA, the specific techniques that proved successful, the logistics of scene

Figure 2.1 The training process is crucial to developing a competent TECT analyst. The trainee must learn all the aspects of the trace collection process and be able to perform under various scene conditions. Collecting trace evidence from the decedent is different in every case.

response, and even manpower issues. The integration of pathology, laboratory expertise, and investigations with the experience of the TECT is one of the most unique features of this team and allows for continual strengthening of the scientific aspects of DNA collection from decedents.

Scene Interactions

TECT analysts collect evidence on scene to identify and protect it before the body is moved (Figure 2.6). During transportation, blood patterns and fluid stains may change or be obscured, and touch DNA may be mixed with contaminating bodily fluids, making it much more difficult to identify the source of foreign DNA (Figure 2.7).

TECT analysts limit collection to the decedent and to objects, such as clothing and bindings, in contact with the decedent. Crime scene personnel from the investigating law enforcement may not participate in collection from the body, and in turn, TECT analysts are not authorized to collect evidence that is not associated with the decedent.

The TECT goal is to assist the law enforcement agency investigating the homicide. Accordingly, the TECT works closely with homicide investigators, sharing information along the way. Although supervised on scene by an HCIFS investigator, their activities are always guided by scientific principles, based on circumstances and scene observations.

Training

Figure 2.2 The observation process is a critical component of the training process. The trainer must ensure that key elements of a collection are conveyed properly to the trainees, so they may apply their knowledge in future cases.

Once the trace evidence is collected, packaged, and labeled, all collected material is transferred to the investigating agency for submission to a crime laboratory for analysis.

In some instances, TECT analysts work with additional HCIFS specialists, including pathologists, anthropologists, and entomologists (Figure 2.8), to assist in determining the cause and manner of death, time of death, and the approximate age and sex of the victim.

Supplies

Below is a list of supplies and materials that are used at the crime scene and found in the TECT crime scene cart that is brought to the scene (Figures 2.9 and 2.10). Different materials are used depending on the crime scene scenario, which will be discussed in detail in the following chapters.

- Forms:
  - Evidence submission forms (Figure 2.11)
  - Trace collection forms (Figure 2.12)
- Personal protective equipment:
  - Tyvek® suit (Figure 2.13)
  - Face masks

8   Forensic DNA Collection at Death Scenes: A Pictorial Guide

Figure 2.3 Once the trainee has completed all required written and oral training exercises, he or she will begin attending crime scenes. The trainee will observe a qualified TECT analyst, and when ready, he or she will begin collecting under supervision. Once the trainee and TECT lead are certain the trainee is ready, a final oral exam involving many different case scenario photographs will be administered. Once all training is complete, the analyst will begin independent TECT collections.

- Hair nets
- Shoe covers
- Gloves (Figure 2.14)
- Sweatbands (Figure 2.15)
- Sweat glove liners (Figure 2.16)
- Lab coats
- Aprons (Figure 2.17)
- Goggles (Figure 2.18)
- Headlamps (Figure 2.19)
- Collection equipment:
  - Evidence pouches
  - Evidence bags
  - Paper envelopes (Figure 2.20)
  - Evidence tape (Figure 2.21)
  - Distilled water (Figure 2.22)
  - Tweezers
  - Swabs (Figure 2.23)
  - Swab boxes (Figure 2.24)
  - Tape lifts (Figure 2.25)

Training

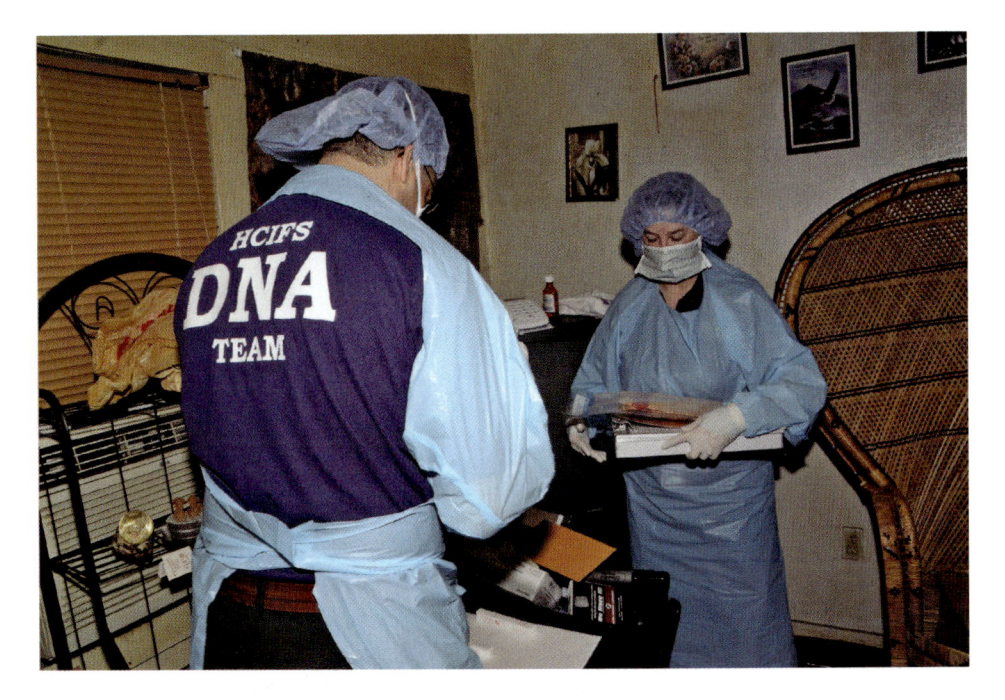

Figure 2.4 Once the trainee has completed all required written and oral training exercises, he or she will begin attending crime scenes. The trainee will observe a qualified TECT analyst, and when ready, he or she will begin collecting under supervision. Once the trainee and TECT lead are certain the trainee is ready, an oral final exam involving many different case scenario photographs will be administered. Once all training is complete, the analyst will begin independent TECT collections.

- Bench paper
- Measuring tape
- Alternate light source/white light
- Rulers
- Gauze pads
- Scene screen (Figure 2.26)
- Crime scene tent

Documentation

Proper documentation is crucial to successful evidence collection. In addition to documenting what was collected and how it was collected, it is also important to describe interventions that took place before collection began since intervention may be a source of contamination. Interventions should be documented in the worksheet. Examples of intervention are the placement of a sheet over the body to protect it from public view. A sheet can cause loss of evidence or contamination of the body and should be documented. Emergency medical technicians (Figure 2.27) may intervene by placing leads on the body during resuscitation efforts. Leads must be documented on the worksheet. Most emergency personnel wear personal protective equipment, which can help limit contamination, but documentation should note any intervention prior to evidence collection.

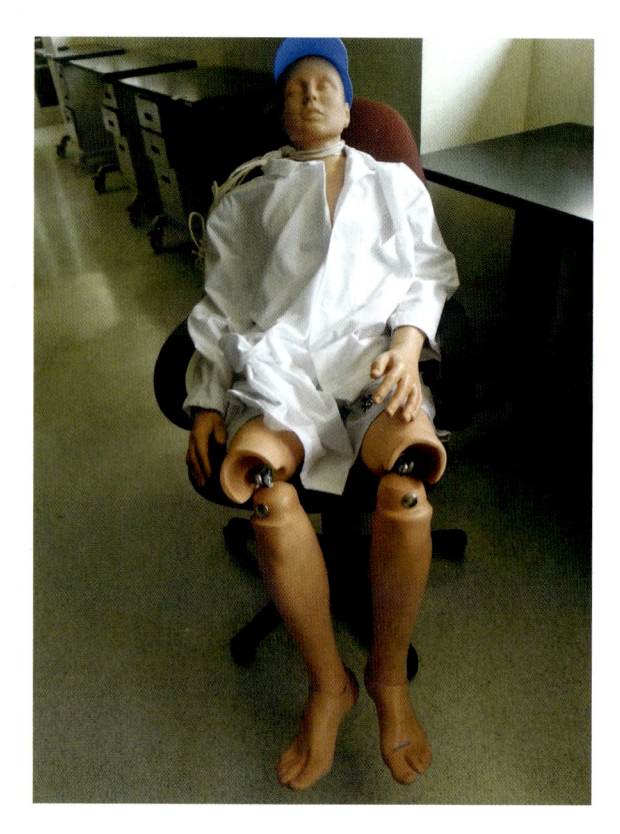

Figure 2.5 Proficiency testing is performed on a latex mannequin that is used to simulate a variety of crime scene scenarios.

Collection Techniques

Before any collection begins, the TECT analyst must be fully gowned. Full gowning includes a plastic apron, shoe covers, a hair cover, a mask, and gloves. A Tyvek suit with a mask and gloves is another option for full gowning. Gowning protects the evidence from contamination and protects the analyst from biological hazards.

Many collection techniques can be utilized at a crime scene. Perhaps the simplest technique for a dry area of the body is a tape lift (Figures 2.28 and 2.29). The sticky side of a labeled tape lift is pressed on the surface repeatedly until the tape is no longer sticky enough to collect material. A tape lift will pick up hair, fiber, debris, and cells containing DNA. DNA can be recovered from tape lifts by swabbing with water or one of several solvents (Figure 2.30). Xylene, for example, can be used to remove adhesive and cells from the tape lift prior to DNA extraction (May and Thomson, 2009). Another technique is to cut tape lifts into many small pieces, which are then extracted in a small tube in the usual manner (Kenna et al., 2011).

If the body is wet, tape lifting cannot be utilized. However, the body can be searched for macroscopic evidence with a white light flashlight or headlamp, and evidence can be collected with tweezers (Figures 2.31 to 2.33), a technique referred to as *picking*. Picked macroscopic evidence can be placed on a tape lift for preservation (Figures 2.34 to 2.36). Commonly, hair, fiber, paint chips, glass, and debris may be collected this way and stored

Training 11

Figure 2.6 Prior to moving a body, body fluids have settled, reducing the possibility of body fluids contaminating foreign DNA that may be present. The skin is still intact, and the body is in the position in which it was discovered. Moving the body can release body fluids.

Figure 2.7 Once the body is transported, body fluids can contaminate areas that may have touch DNA. If the body is slightly decomposed, skin slippage occurs. Trace evidence can move with it.

Figure 2.8 The TECT works in concert with the pathologists, investigators from the medical examiner's office, investigators from the law enforcement agency, entomologists, and anthropologists. The TECT team can assist both the death investigators and the law enforcement investigators.

Figure 2.9 The TECT rolling cart houses the supplies needed to process and package evidence at a crime scene.

Training

Figure 2.10 The TECT rolling cart is used to bring supplies to the crime scene. This cart keeps all supplies organized and easy to find. The cart can roll on any terrain and can be maneuvered up stairs as well. The cart also doubles as a clean station to collect and package evidence.

on a labeled tape lift. The tape lift is placed in a labeled paper envelope until transfer to the investigating agency (Figure 2.37).

Swabbing is another useful technique. Note that DNA from cells in saliva has been reported to remain on human skin up to 96 hours after transfer by licking (Kenna et al., 2011). This saliva and touch DNA can be effectively recovered by swabbing with one wet swab followed by one dry swab (Figure 2.38). The first swab is moistened with deionized water and excess water is flung from the swab. The target area is then swabbed while twisting the swab to ensure that the collected material is uniformly distributed around the head

TRACE EVIDENCE COLLECTION SUBMISSION FORM
Harris County Institute of Forensic Sciences
1885 Old Spanish Trail, Houston, Texas 77054
www.co.harris.tx.us/ifs
Main: (713) 796-9292
Fax: (713) 796-6794

ML/OC #: _____   Police Agency: _____

                             Police Agency #: _____

Evidence Collected: ___/___/___   Time: _____

Evidence Collected by: _____   _____
                       Analyst's Name   Analyst's Signature

Outside Container: _____

Item #	Qty.	Full Description of Evidence

Drop Box _____   Hand Delivered _____

Date Released: ___/___/___   By: _____   Released to/Time: _____
Date Released: ___/___/___   By: _____   Released to/Time: _____
Date Released: ___/___/___   By: _____   Released to/Time: _____

White Copy – Laboratory   Yellow Copy – Receiving Officer   Pink Copy – Case File

Form #: TCTF08.015   Rev.: 3   Procedure #: TCT08.2001

Figure 2.11 The TECT submission form is used to release evidence to law enforcement investigators at the scene and also to keep the chain of custody intact. The form is in three parts, an original and two carbon copies. One copy remains with the trace team, one is given to the officer at the scene, and the original remains with the evidence.

of the swab. Then a dry swab is used to collect the remaining water and DNA that may be left behind on the surface. This swabbing technique is ideal for nonporous materials, such as metal, glass, and plastic (Williamson, 2012). If the swabbed area is already wet, there is no need to moisten. In some instances, the swabbed surface is porous. In that case, a dry swab is not used because residual water will not remain and two wet swabs can be used.

When swabbing the skin of the decedent, take care to exert only medium pressure to maximize the collection of epithelial cells from the suspect while minimizing collection of

Training

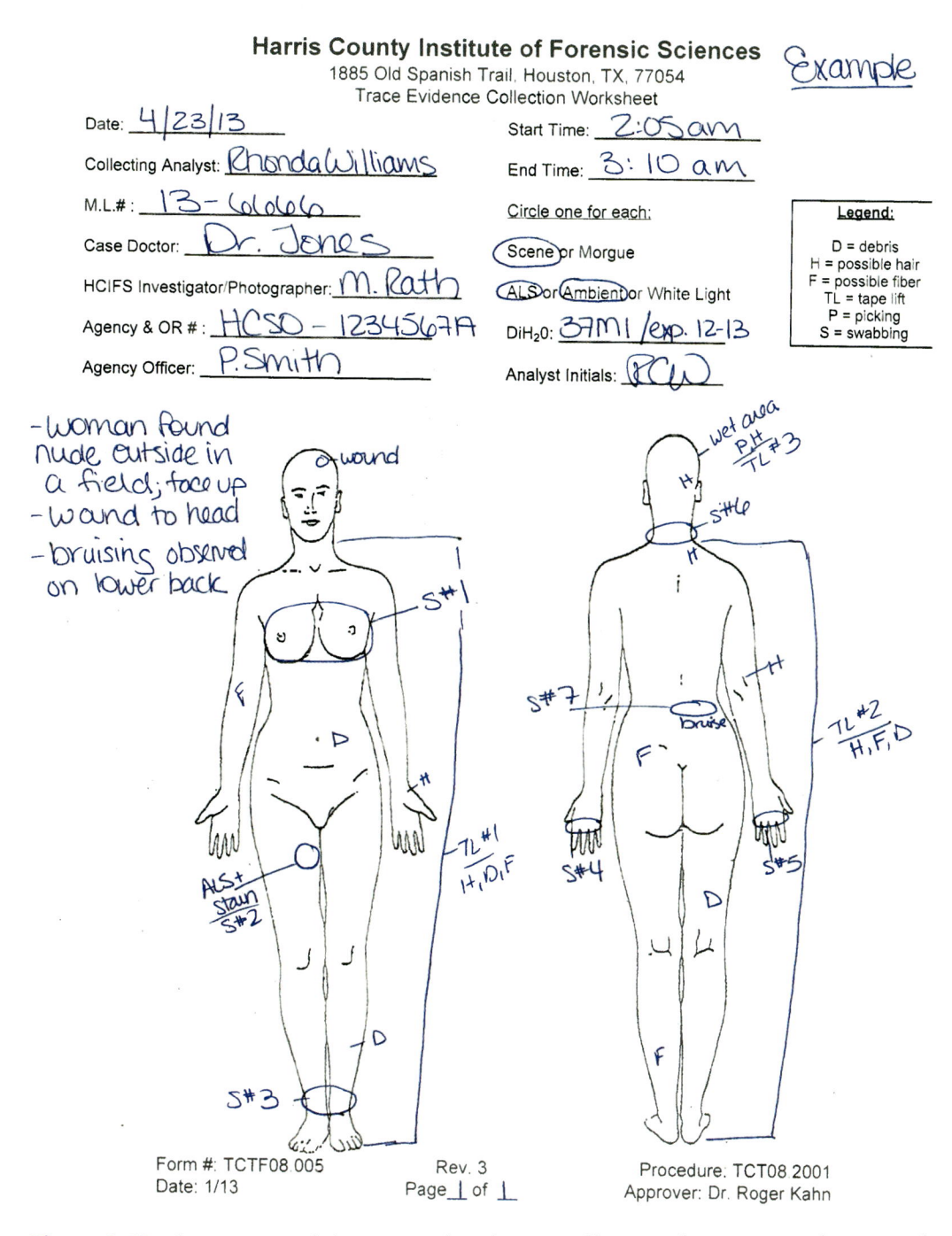

Figure 2.12 The TECT worksheet is used to diagram all areas where trace evidence is collected. Tape lifts, swabs, and pickings will be listed with the location of collection on the body. The scene scenario is also placed on the worksheet with the case numbers and TECT analyst's name. The date, the time, and the lot number of the DNA-free water used are also listed. The worksheet summarizes the collection, and it is given to the pathologist and the detective in charge of the case after review by the trace team lead.

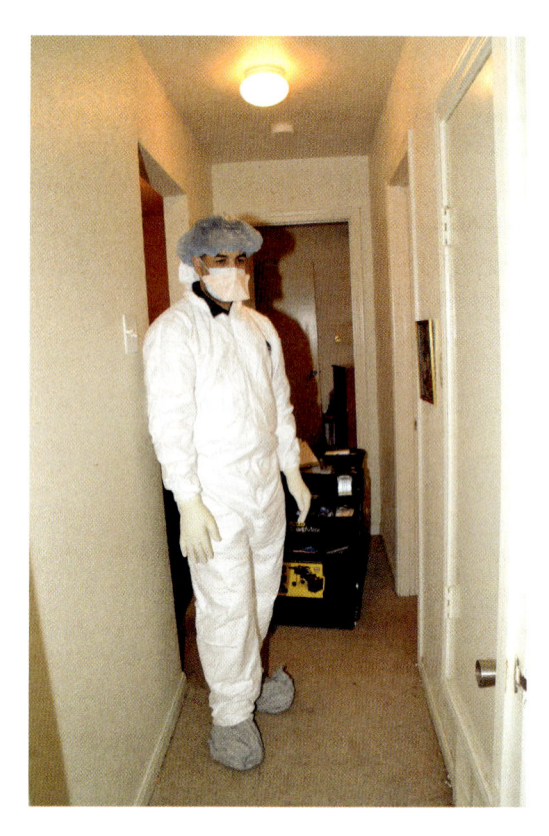

Figure 2.13 Tyvek suits are used when scenes are very messy or when the analyst is required to crawl or climb through dirty areas. This suit encloses the analyst from head to foot except for the face, where a mask is used, and the hands, where gloves are used.

Figure 2.14 Gloves are used at the crime scene and are changed frequently. The swabs and tape lifts should be considered separate items that must not contaminate one another. Gloves should be changed when they touch a surface that could contaminate another surface.

Training 17

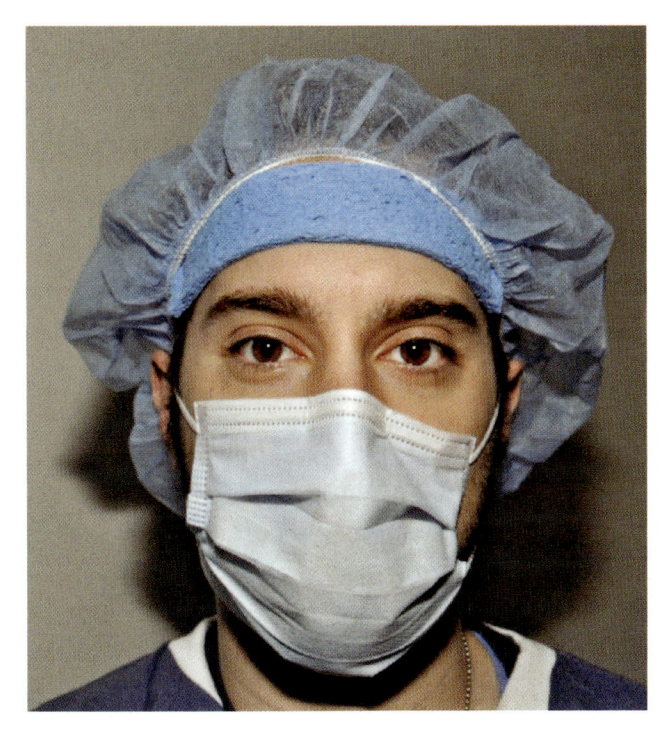

Figure 2.15 Sweatbands keep sweat from dripping on the evidence causing contamination. Sweatbands also keep sweat out of the analyst's eyes. With a band in place, the chance of sweat causing contamination is reduced.

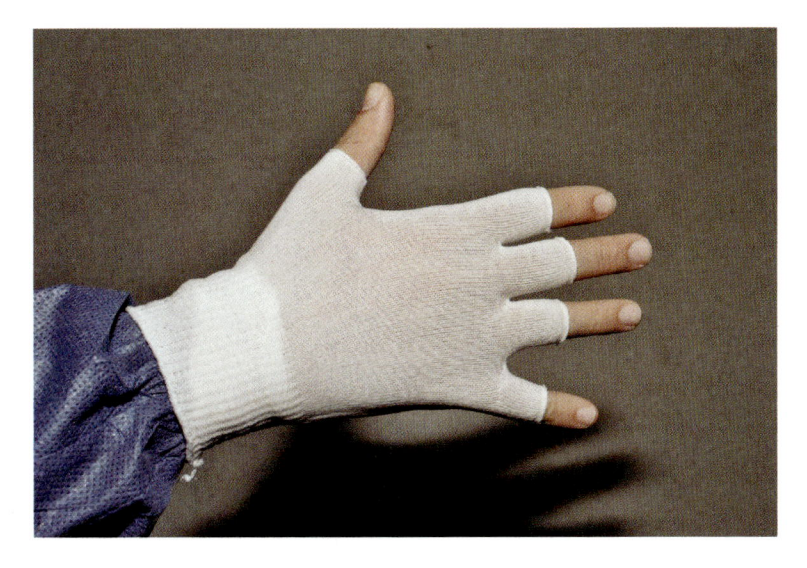

Figure 2.16 Glove liners are worn beneath the gloves to ensure that sweat stays contained and does not come out of the gloves with each change. The liners also allow an easy glove change because the hand is dry. Sweat droplets can cause contamination and must be kept away from surfaces and samples during collection.

Figure 2.17 The TECT analyst can choose to use an apron if the scene is not messy enough to require a Tyvek suit. The apron is plastic and waterproof to ensure body fluids do not come in contact with the analyst or his or her uniform.

skin from the decedent. Bloody areas of the body are generally avoided as target sites for touch DNA collection.

Swabs are packaged in a labeled swab box with the hole punched out to encourage rapid air drying (Figure 2.39). The swab box is placed into a labeled paper envelope, ready for submission to the investigating agency (Figure 2.40).

Swabbing may not be practical when an extensive surface must be sampled. In this instance, gauze may be used instead (Figure 2.41). The gauze may be moistened with deionized water and used to wipe the item or area. Afterward, the gauze can be placed in its original sleeve with the collection side marked with a dot for identification. The sleeve can then be placed in a labeled paper envelope and submitted to the investigative agency (Figures 2.42 and 2.43).

Alternate light sources can assist the analyst in determining where to swab the body. Blue wavelength light (420–470 nm) viewed through orange safety goggles will fluoresce where semen or saliva may be present (Saferstein, 2009). This light can be used if the scene is dark enough to view the fluorescence. Black pop-up tents can be brought to the scene to darken the area when an alternate light source is necessary in daylight conditions. Tents are also very useful if scene conditions, e.g., rain or high winds, could lead to loss of DNA.

A variety of items may be used to make trace collections easier. We have found a rolling trace cart, headlamps, sweatbands, glove liners, and scene screens are all helpful. The

Training

Figure 2.18 Orange goggles are used as a safety precaution with a blue wavelength alternate light source. The light causes semen and saliva to fluoresce, helping to identify the location of the fluids when they are present on the body. This light/goggle combination can only be used when the scene is dark or can be darkened to view the fluorescence.

trace cart we use is sturdy and durable. It can be rolled over nearly any terrain and is easy to maneuver, even up stairs. The cart has many divided sections to store equipment, and it has a top lid that folds over and doubles as a workstation for the analyst.

Headlamps, worn around the forehead, free the hands and make collection easier and more effective. Disposable sweatbands are worn around the hairline to avoid contamination from sweat while working in Houston heat and humidity. Sweatbands are an excellent way to keep sweat from dripping on valuable evidence while also keeping sweat out of the analyst's eyes. Glove liners prevent sweat from dripping out of gloves when collecting evidence, and they allow the analyst to change gloves more easily by providing a dry layer between the skin and the glove. Lastly, a scene screen is an excellent way to prevent bystanders from viewing the body and the scene, or from disturbing the analyst. The screen also forms a line of demarcation to keep other personnel from the collection area. Crowding over the body is a common problem that can cause distraction and contamination. A screen can help to eliminate this problem. All of these items have been very useful in the collection process and are must-haves for any analyst collecting at a crime scene.

These standard techniques can be used in many different scenarios to recover valuable evidence that can assist in an investigation. Tape lifts will collect macroscopic evidence, as well as DNA. Swabbing and gauze swabbing are best for collecting touch DNA from the decedent.

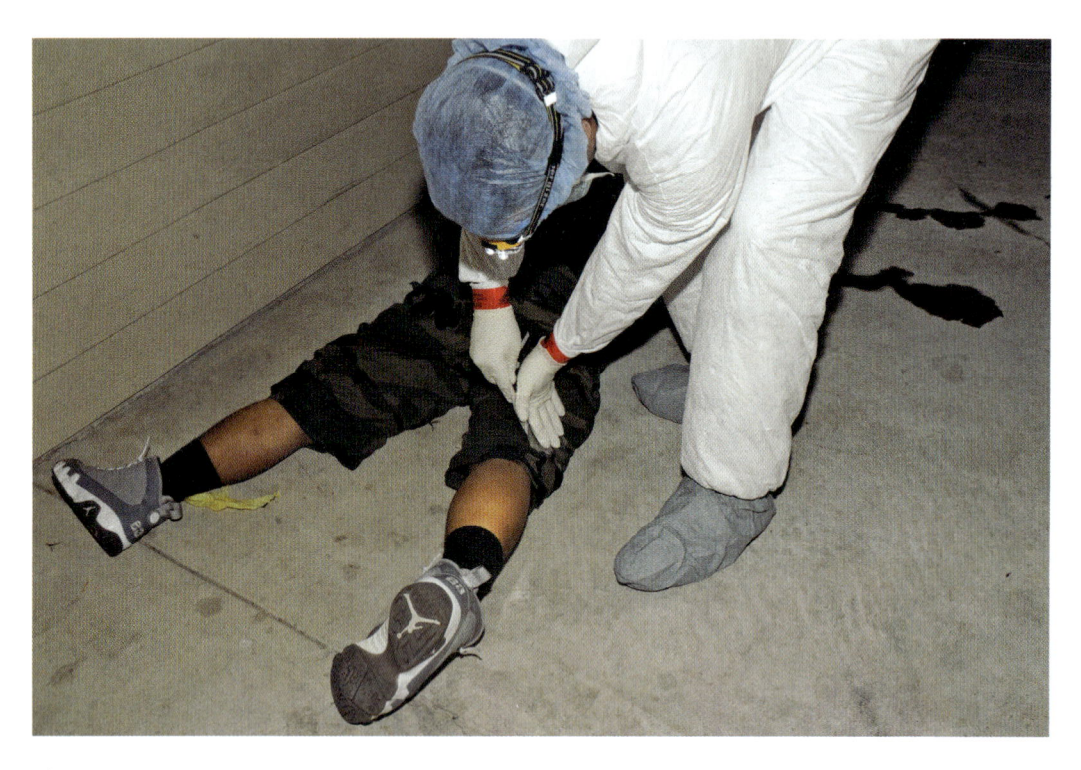

Figure 2.19 Headlamps can be used at scenes where it is dim or dark. The headlamp leaves both hands free during the collection.

Training 21

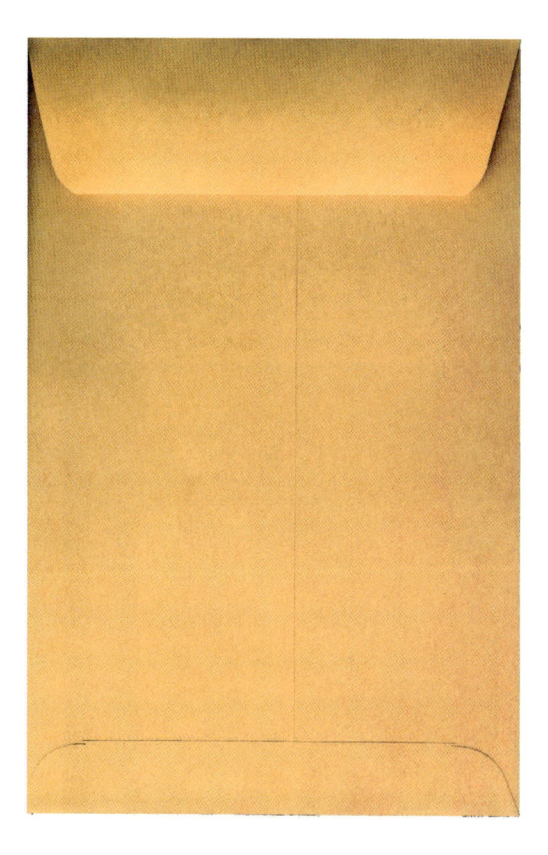

Figure 2.20 Paper envelopes are used to store swab boxes, gauze, or tape lifts. Paper is the preferred packaging, as it allows airflow that that can inhibit or prevent mold by allowing the sample to dry. The paper envelope stores a subitem to protect it from cross-contamination with other items. The paper envelope is labeled with the case number, item number, collecting analyst, and date. Multiple swab boxes can be placed in one paper envelope. The swab box protects against cross-contamination, unlike tape lifts, which are packaged separately.

Figure 2.21 Tamper-evident evidence tape is used to seal paper envelopes. The collecting analyst adds initials and date on the tape.

Figure 2.22 Deionized water, which is purchased DNA-free, is used to moisten swabs prior to collection of evidence in a dry location. If DNA-free water is not available, clean water or saline can be used instead. If the area of collection is already wet, e.g., with blood, there is no need to moisten the swab. In some instances a dry swab is used after the wet swab to collect sample left behind.

Training 23

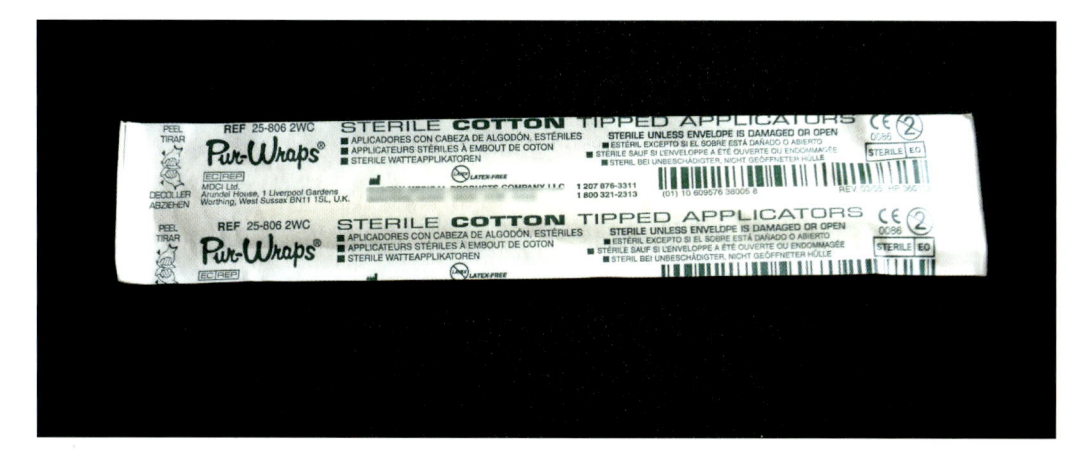

Figure 2.23 Swabs are used to collect evidence at the crime scene. If the evidence is wet, deionized water is not needed to moisten the swab. If the evidence is dry, the swab is moistened prior to collection. It is important to rotate the swab with your fingers while collecting to ensure the sample is evenly distributed on the swab. If residual water is left behind, a dry swab can follow the wet swab to ensure that nothing is left behind. The swabs are then placed within a labeled swab box and then a labeled paper envelope.

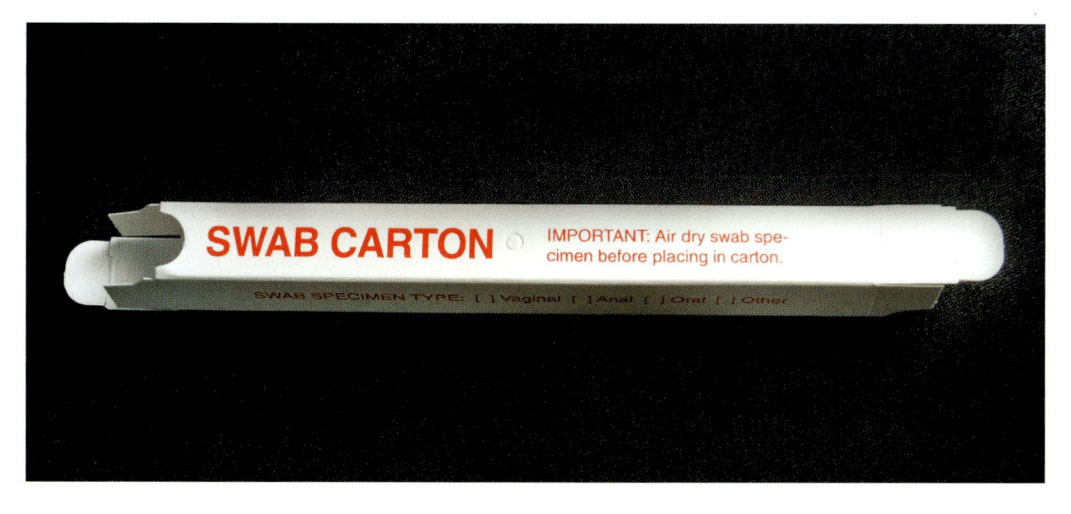

Figure 2.24 Swab boxes are used to store the collected evidence swabs. Holes in the boxes are punched to allow the swabs to air-dry. Swab boxes must be labeled with the case number, item number, item description, collecting analyst's initials, and date. The swab boxes are stored within a labeled, paper envelope. To save space, several swab boxes can be placed in one paper envelope as the box protects against cross-contamination.

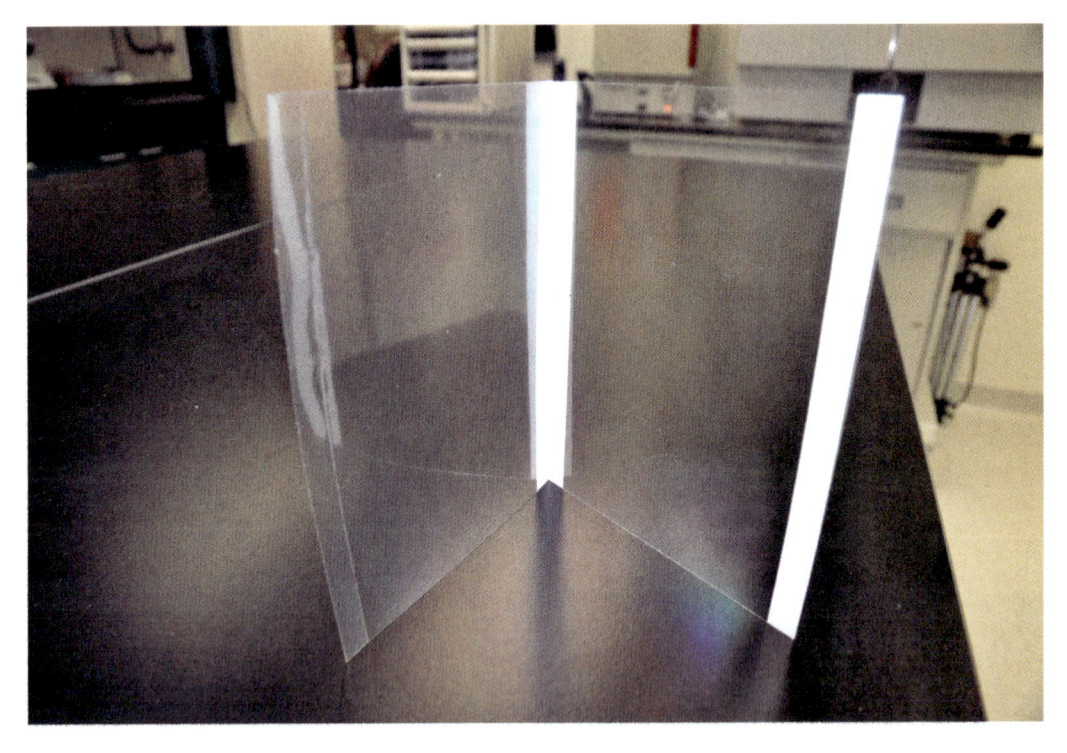

Figure 2.25 Tape lifts are used to collect hairs, fibers, touch DNA, and debris from an object. These lifts fold on themselves once used and stay together because of the adhesive that remains. Tape lifts must be labeled with the case number, item number, item description, collecting analyst's initials, and date. Once the tape lift has been labeled, it is placed in a paper envelope. Only one tape lift is placed in an envelope, as there is a potential for cross-contamination.

Training 25

Figure 2.26 The scene screen can be used if the scene is outdoors in public view. It creates a visual barrier between the scene and the public. It is also useful for ensuring the collecting analyst has enough room to work. Crowding over the body can be an issue on occasion, and contamination can be a concern. The screen helps to mitigate the problem.

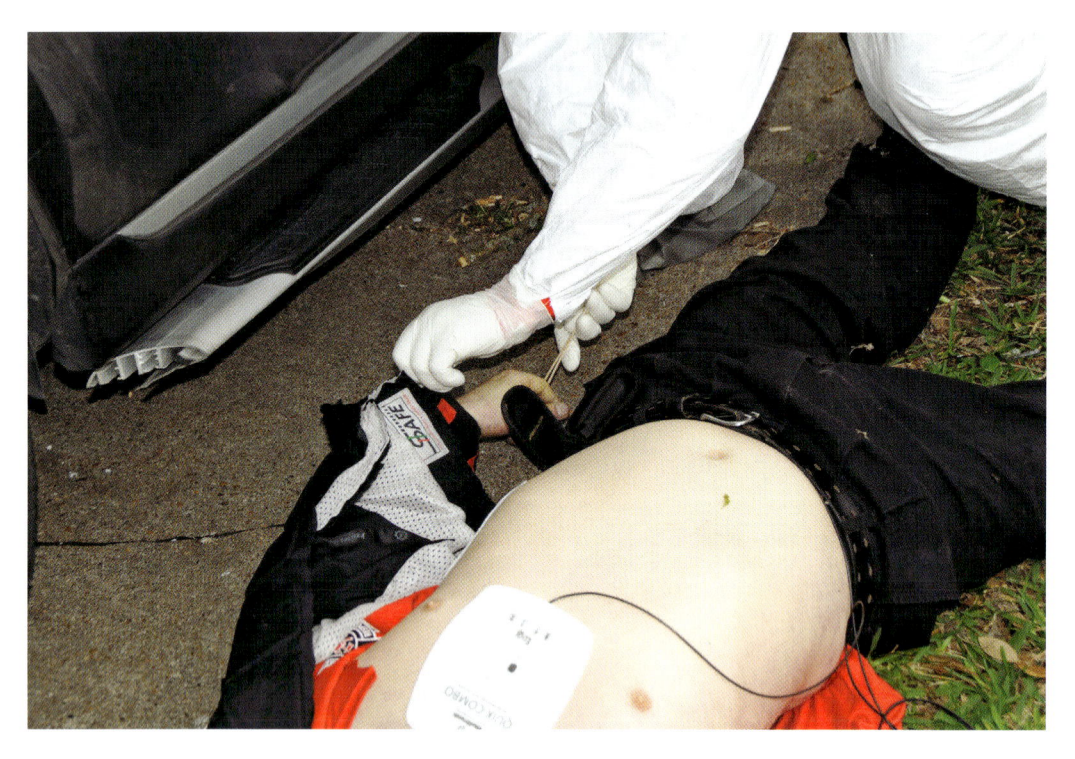

Figure 2.27 Medical intervention prior to the trace collection must be documented on the worksheet. This provides a written record of sources of possible contamination prior to collection.

Training

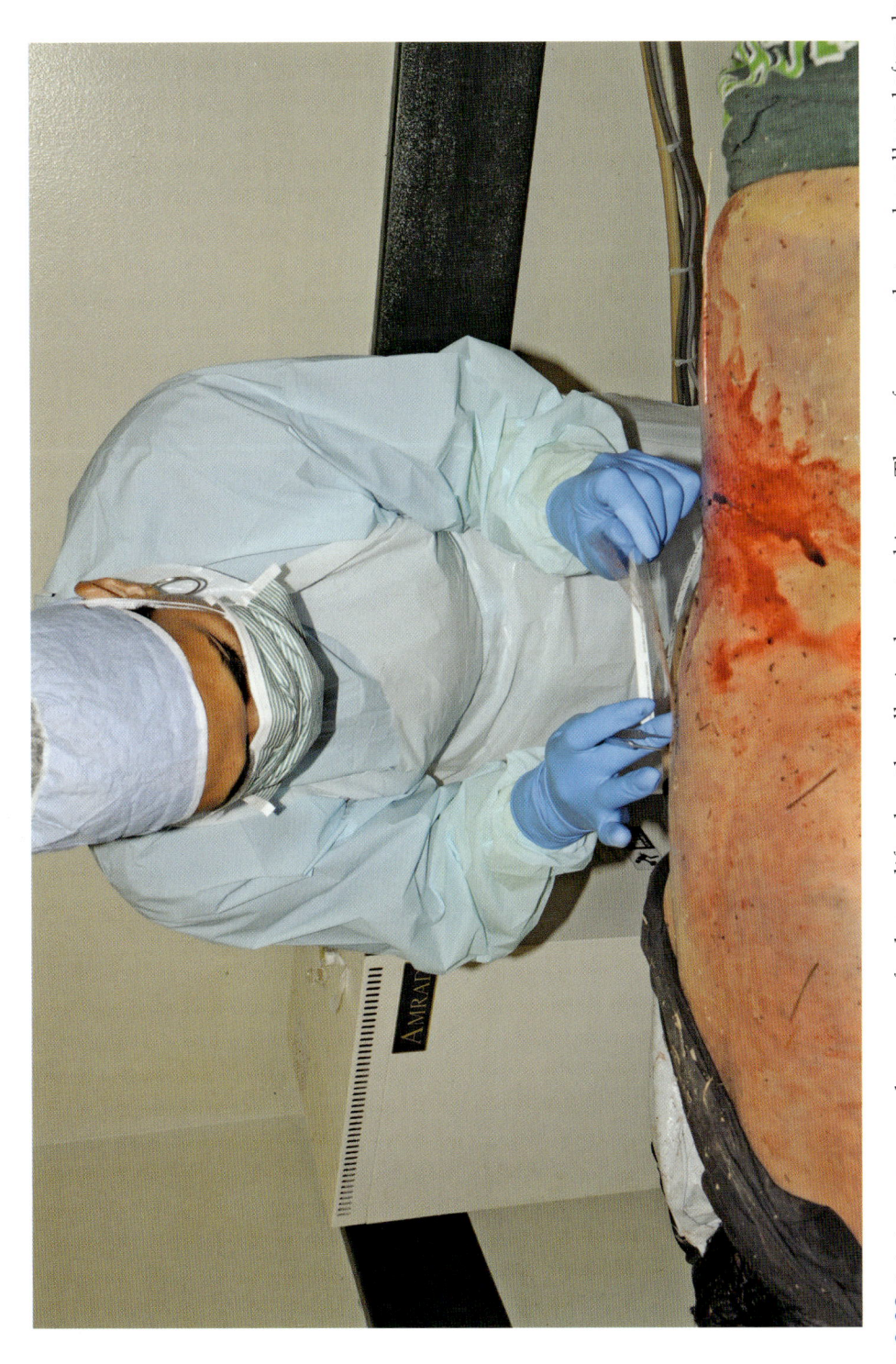

Figure 2.28 It is important to change to a fresh tape lift when the adhesive loses tackiness. The surface area that can be collected often depends on the amount of debris on the area of collection. The lift must be pressed in a uniform manner to ensure that it comes in contact with all crevices of the surface.

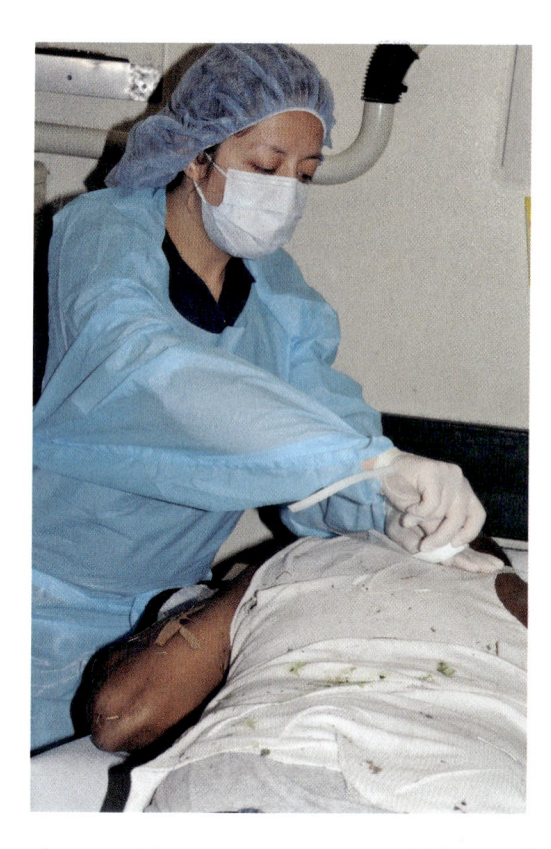

Figure 2.29 Tape lifts of areas with excessive amounts of debris will cover less surface area than a cleaner surface. If the area to be tape lifted is wet, items of importance should be picked and placed on the tape lift for preservation.

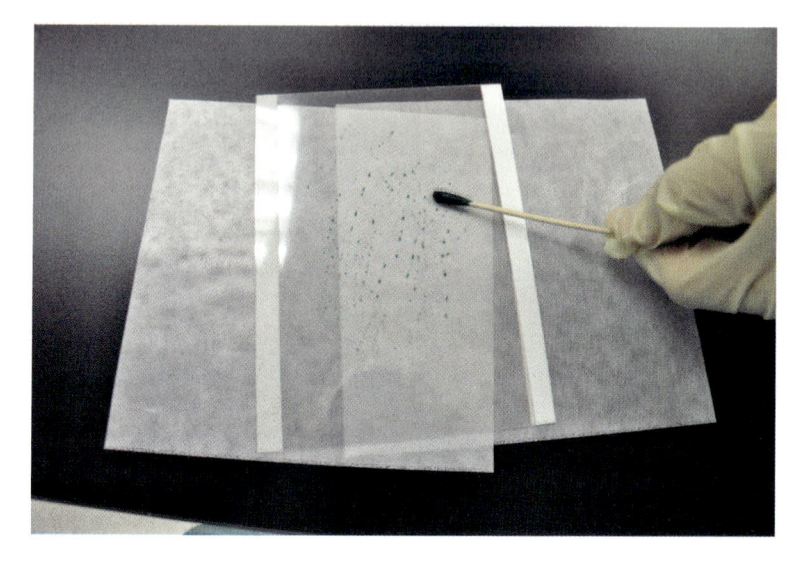

Figure 2.30 Tape lifts are normally used to collect hair and fibers from clothing and from the body. Tape lifts also collect epithelial cells for touch DNA. Tape lifts can be swabbed with water or xylene to retrieve the touch DNA present.

Training

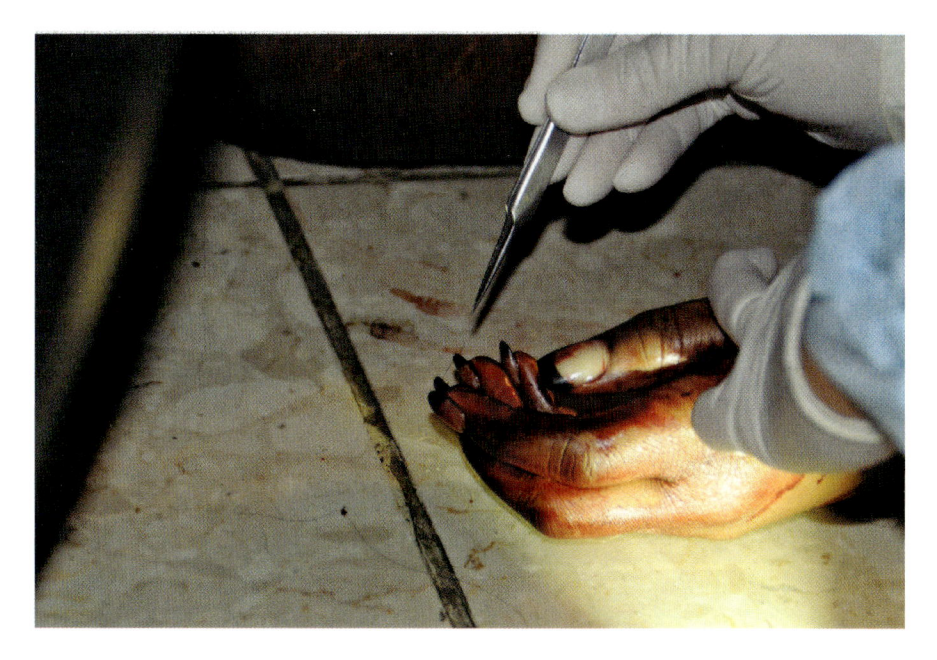

Figure 2.31 Tweezers are used when the area is too wet to be tape lifted. They can also be used to pick up macroscopic particles such as glass, hairs, and paint chips. It is important to photograph this collection to identify where the evidence was located.

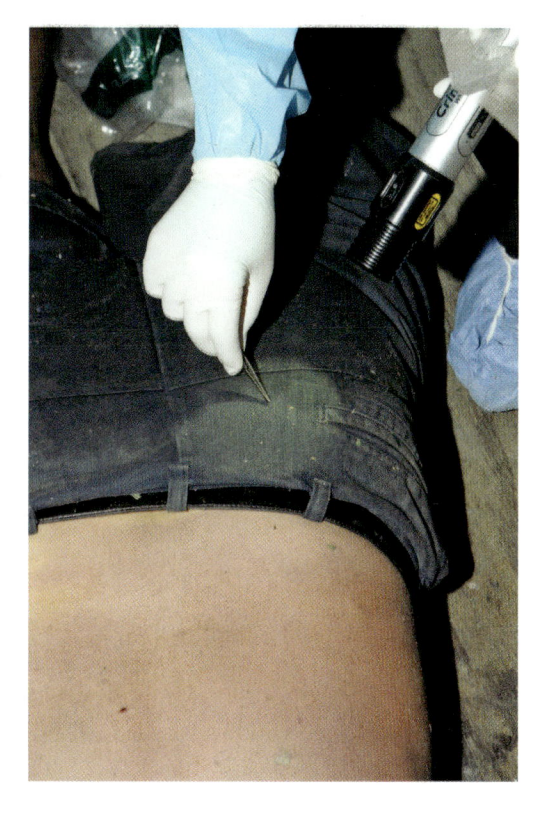

Figure 2.32 If the body is wet, tweezers must be used to collect evidence. A flashlight can help to identify this evidence. Photographs should be taken of collected evidence.

Figure 2.33 Some types of evidence are stuck on the body and will not be lifted by a tape lift. In these instances, tweezers can be used to help in the collection of these pieces of evidence.

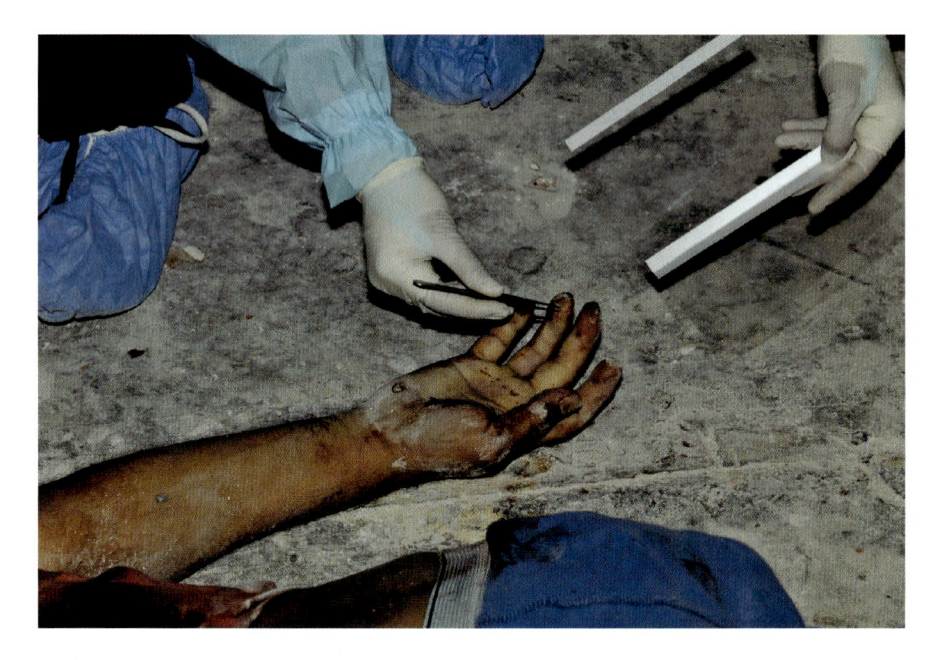

Figure 2.34 When the body is wet, tape lifts cannot be used. Tweezers should be used to remove macroscopic evidence. This evidence should be photographed and documented prior to collection.

Training 31

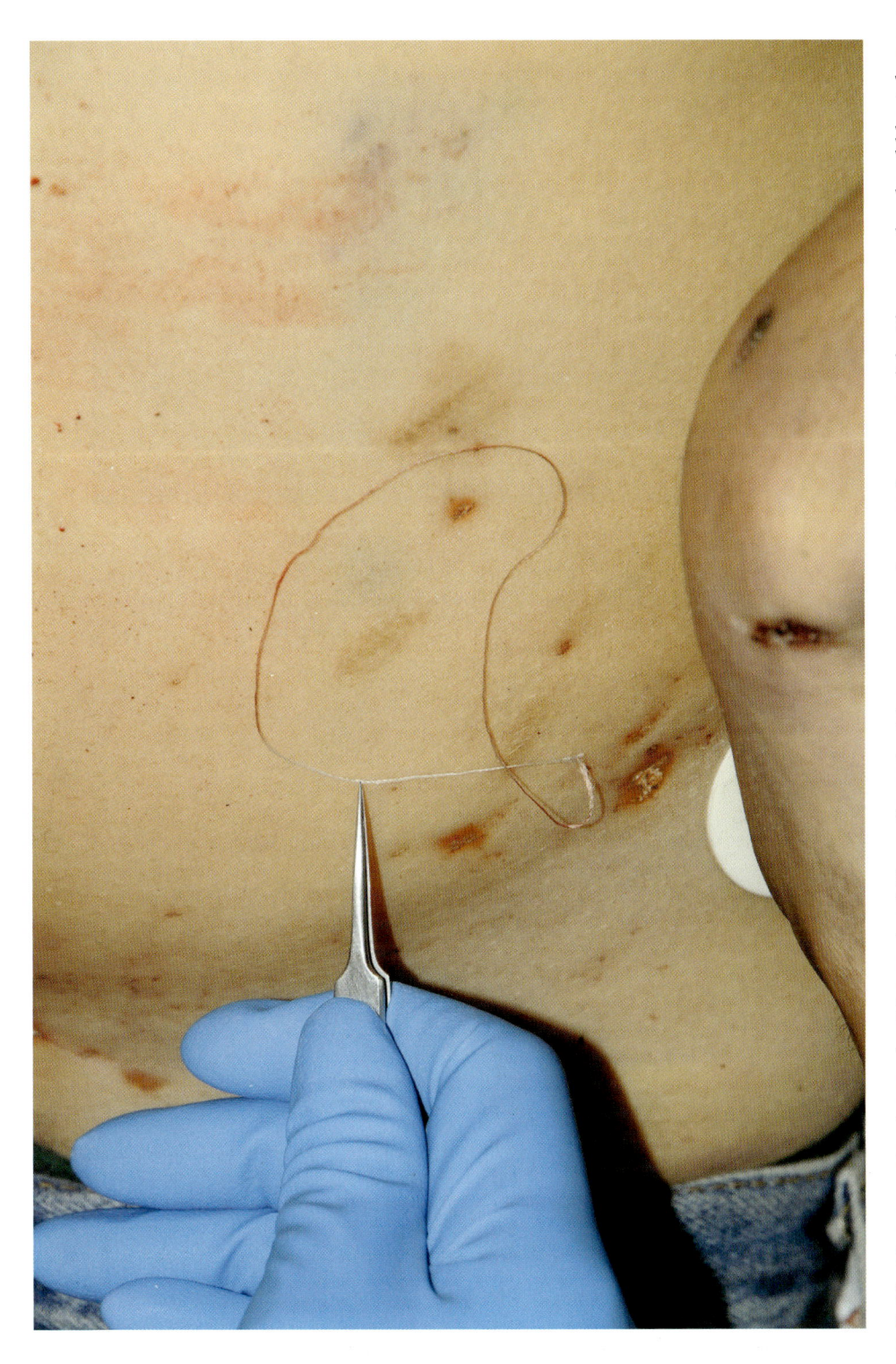

Figure 2.35 Macroscopic pieces of evidence should be collected with tweezers and secured on a tape lift. Photographs should be taken prior to collection.

Figure 2.36 Macroscopic evidence should be collected with tweezers and placed on a tape lift. If the evidence is too bulky for a tape lift, it should be placed within a small, paper envelope. All macroscopic evidence that is removed from the body should be photographed prior to removal.

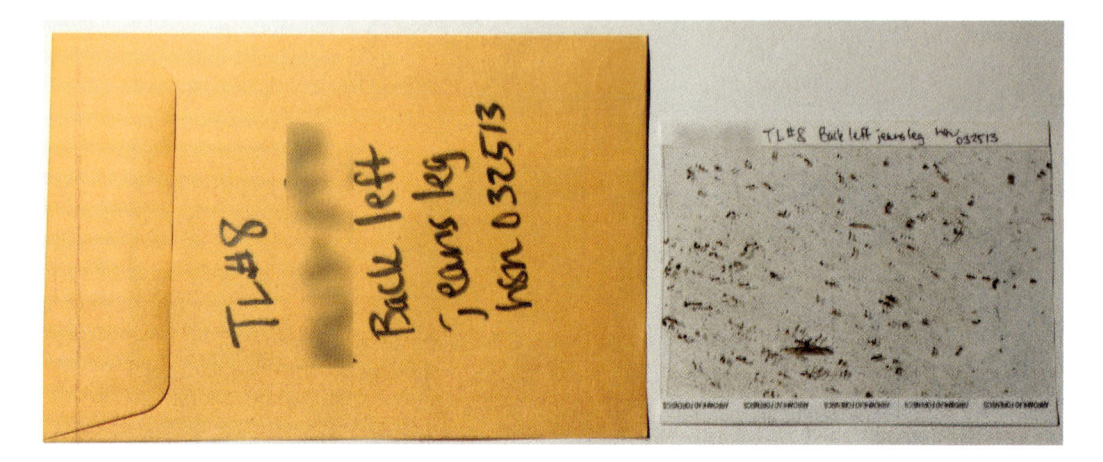

Figure 2.37 Tape lifts should be labeled with the case number, item number, item description, date, and collecting analyst's initials. A picture should be taken to document this evidence, as it is released at the scene to proper law enforcement and is not available for physical review by the team lead. The tape lift will then be placed into a labeled paper envelope and released to law enforcement.

Training 33

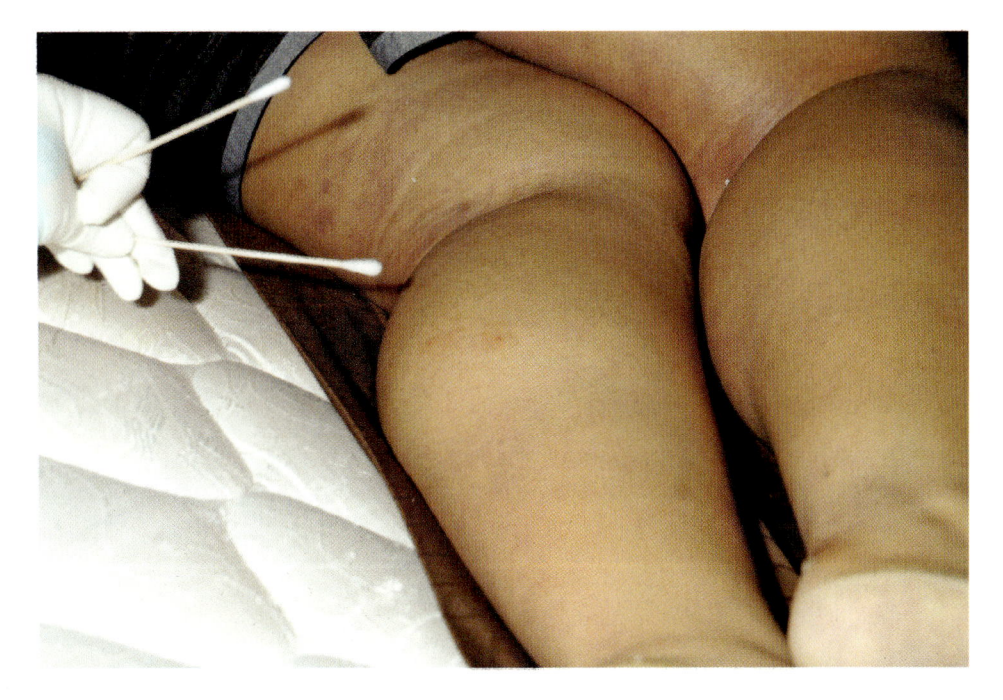

Figure 2.38 Swabbings of nonporous objects should be performed with one wet (damp) swab followed by one dry swab. The dry swab is used to pick up the residual sample that remains.

Figure 2.39 Swab boxes that contain wet swabs should have the perforated hole punched out to allowed air to dry the swab and prevent mold.

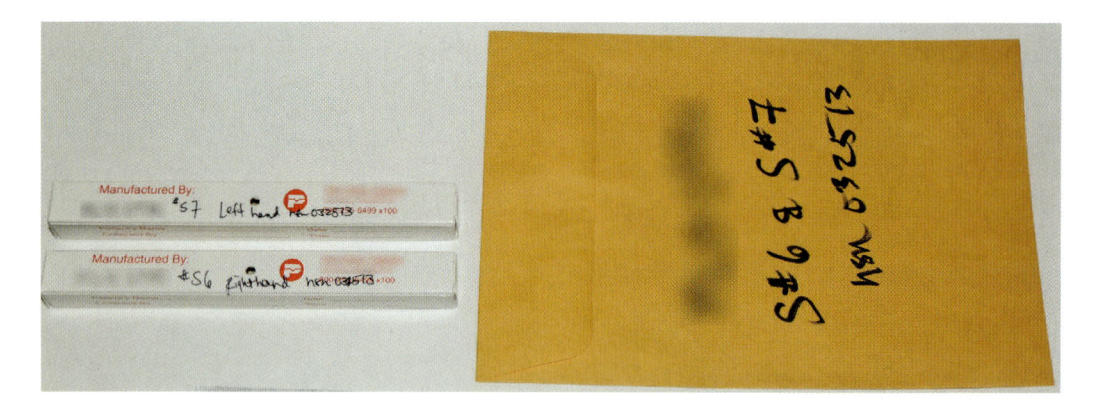

Figure 2.40 Swab boxes should be labeled with the case number, item number, item description, date, and collecting analyst's initials. A photo should be taken to document the evidence before it is released to law enforcement investigators at the scene, as it will not be available for later review.

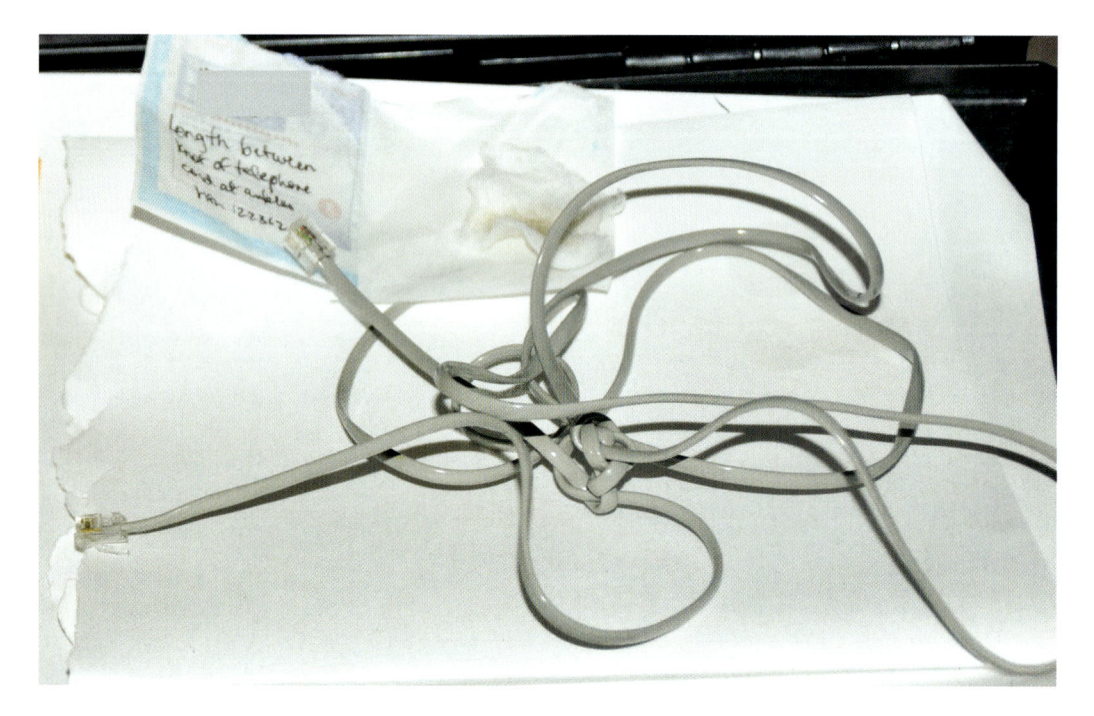

Figure 2.41 Gauze is used to collect touch DNA from large surfaces where swabbing would become cumbersome. The gauze pad is rubbed over the surface and placed back into its original sleeve. The side used for collection is noted with a dot on the gauze. The packaging is labeled and then placed in a labeled, paper envelope.

Training 35

Figure 2.42 The gauze sleeve should be labeled with the case number, item number, item description, date, and collecting analyst's initials. A picture should be taken to document it before it is released at the scene; it will not be available for later review. The gauze in the sleeve is placed in a labeled, paper envelope prior to release.

Figure 2.43 Gauze sleeves should be packaged in individual envelopes to prevent cross-contamination.

Preservation and Packaging 3

Preservation is very important at any crime scene. In some instances, the evidence on the body cannot be removed and packaged (Figure 3.1). If the item is the instrument of death, it cannot be moved prior to the medical examiner's autopsy. A knife in the body must remain in the body during transport to the medical examiner's office or while examined by the pathologist at the scene. The knife must be protected at the scene prior to transport to prevent damage to the knife or additional damage to the body or loss of evidence. The TECT will place a brown paper bag over the knife and tape it to preserve fingerprints and DNA evidence that may be on the knife (Figures 3.2 to 3.5). As another example, a bullet hole through clothing or through bindings must be preserved, if possible.

In one instance, the decedent was found with duct tape around the head and a bullet hole through the duct tape. The duct tape could not be removed and had to be preserved to retain any possible evidence on the tape itself. The TECT analyst moved the head very slightly to analyze the body for trace, and then the head was held still as the body was placed into the bag to prevent blood from flowing onto the tape, resulting in the loss of evidence on it. Careful handling of the body can be critical in a variety of situations.

In some situations, it is not as important to keep the evidence in contact with the body. For example, duct tape found around the hands and feet can be removed and separately preserved for analysis of fingerprints and DNA. The position of the tape, how tightly it is wrapped, and whether it is over clothing, for example, must all be well documented before removal (especially with photographs). It will likely be the pathologist who testifies to details of binding on the body; moreover, such bindings can leave artifactual marks or real injuries on the body. These findings on the body are best interpreted with knowledge of the details of the ligatures. Direct consultation with a forensic pathologist usually occurs before a TECT analyst removes such items (Figure 3.6). The TECT will cut the tape in an area away from the duct tape's ends and place it sticky side down on the shiny side of a piece of laboratory bench paper. The paper is rolled into a cylinder with the tape on the convex aspect, and placed inside a brown paper bag, taking care that the sides do not touch the bag or otherwise damage fingerprints on the tape (Figures 3.7 and 3.8).

All collected items should be placed in paper envelopes to allow the evidence to breathe and dry properly. This prevents the evidence from molding and destroying the DNA or reducing the amount of DNA recovered. Outer evidence packaging should be properly labeled with the case number, item number, collecting analyst's initials, and date (Figure 3.9). Swab boxes and tape lifts must also include the item description. If any of the label information is incorrect, it must be corrected with a single line through the error, initialed, and dated; the correction should be placed near the erroneous information. The correction should be documented with a photograph (Figure 3.10). We use plastic evidence pouches that have a mesh material on one side to allow the evidence to breathe. Paper envelopes are placed inside the plastic pouch to ensure that they do not get wet or tear during transport (Figure 3.11).

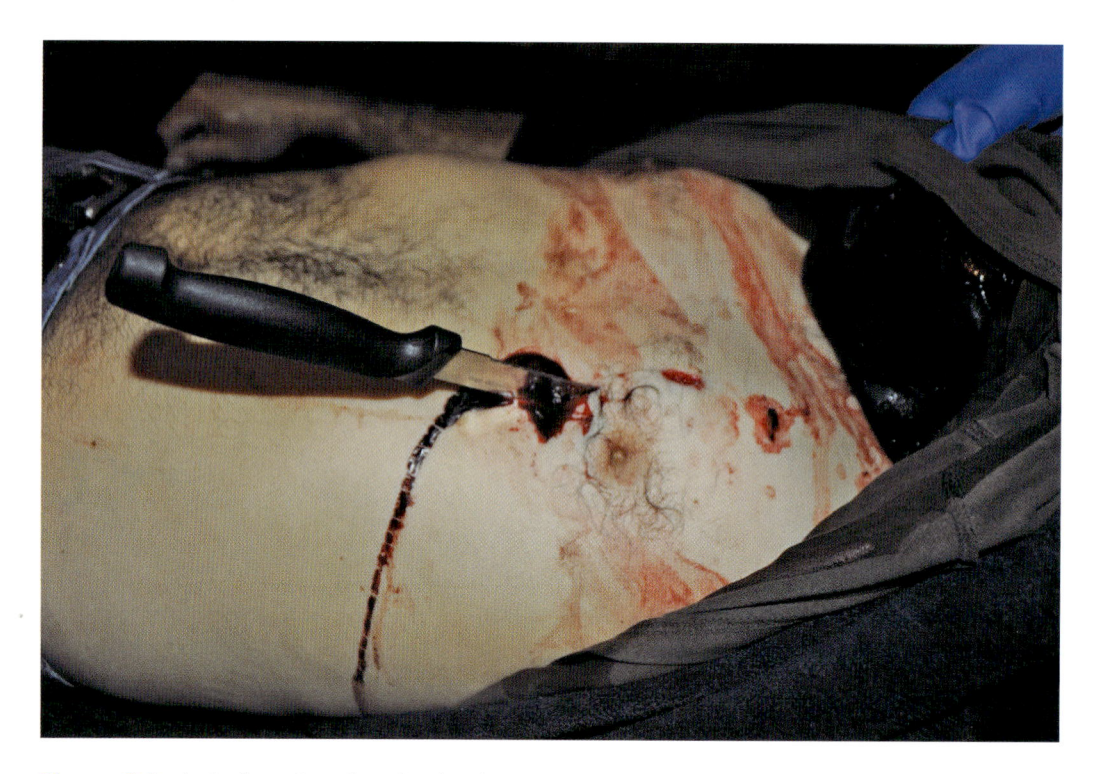

Figure 3.1 If a body is found with a knife inside, the knife should remain there.

When a collected item is larger than a paper envelope, it should be placed inside a paper bag (Figure 3.12), and the bag is labeled in the usual manner. If the item is wet, it must be dried prior to placing it in the bag. If the item must be transported wet, it can be briefly placed within a plastic bag for transport, allowed to dry outside the bag, and then transferred to a paper bag once it is dry (Figure 3.13). Other items that may be preserved for transport are pockets (Figures 3.14 and 3.15), clothing (Figures 3.16 and 3.17), hands (Figures 3.18 and 3.19), and bloodstain patterns on clothing (Figures 3.20 and 3.21).

Preservation and Packaging

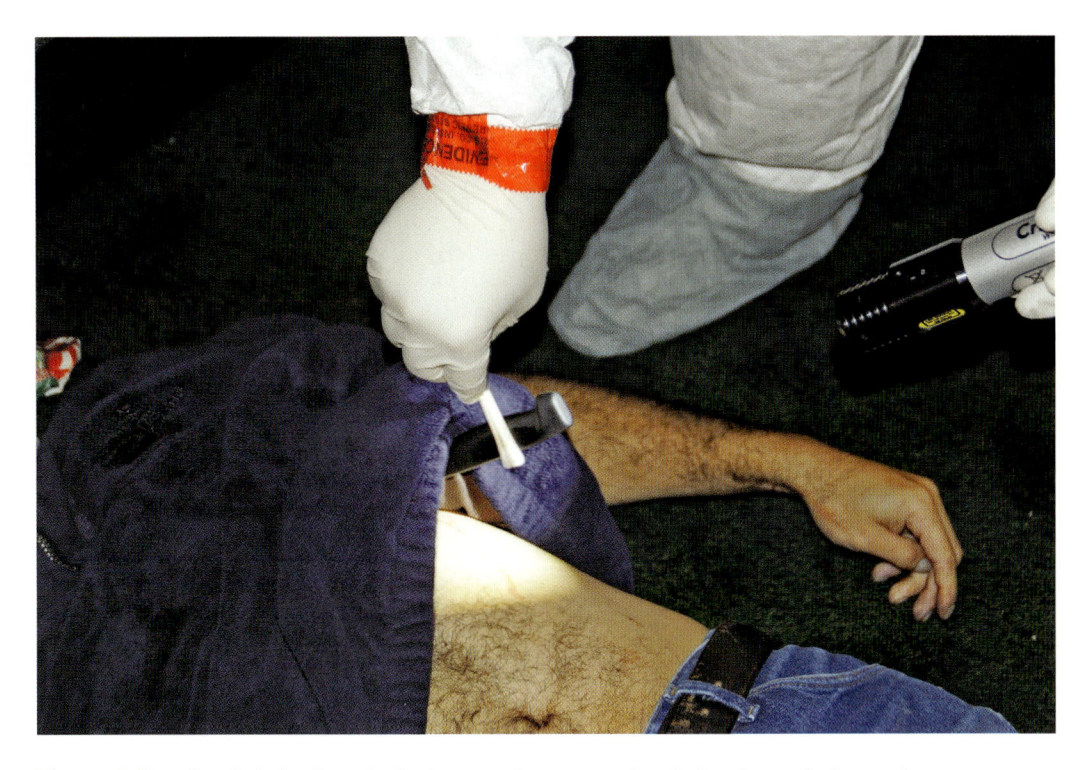

Figure 3.2 A knife lodged in the body must be removed only by the pathologist during autopsy. If there are no visible fingerprints, and law enforcement at the scene is not requesting prints, the knife handle can be swabbed for DNA prior to transporting the decedent to the morgue. If blood is present on the handle, avoid it and swab only blood-free areas. Once the knife handle is swabbed, protect it with a paper bag. A more thorough collection can be done in the DNA laboratory after the knife is submitted.

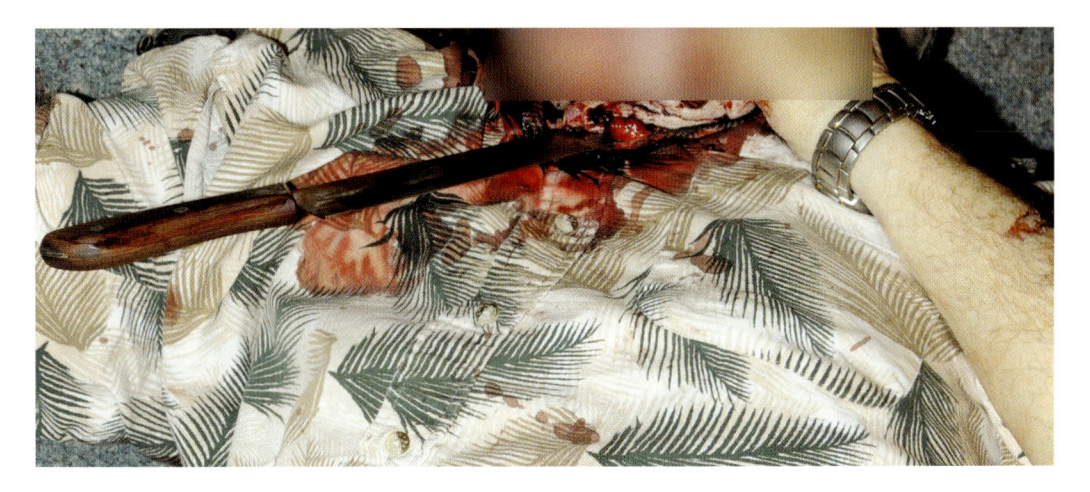

Figure 3.3 A knife lodged in the body should be preserved for fingerprint analysis and subsequent DNA analysis.

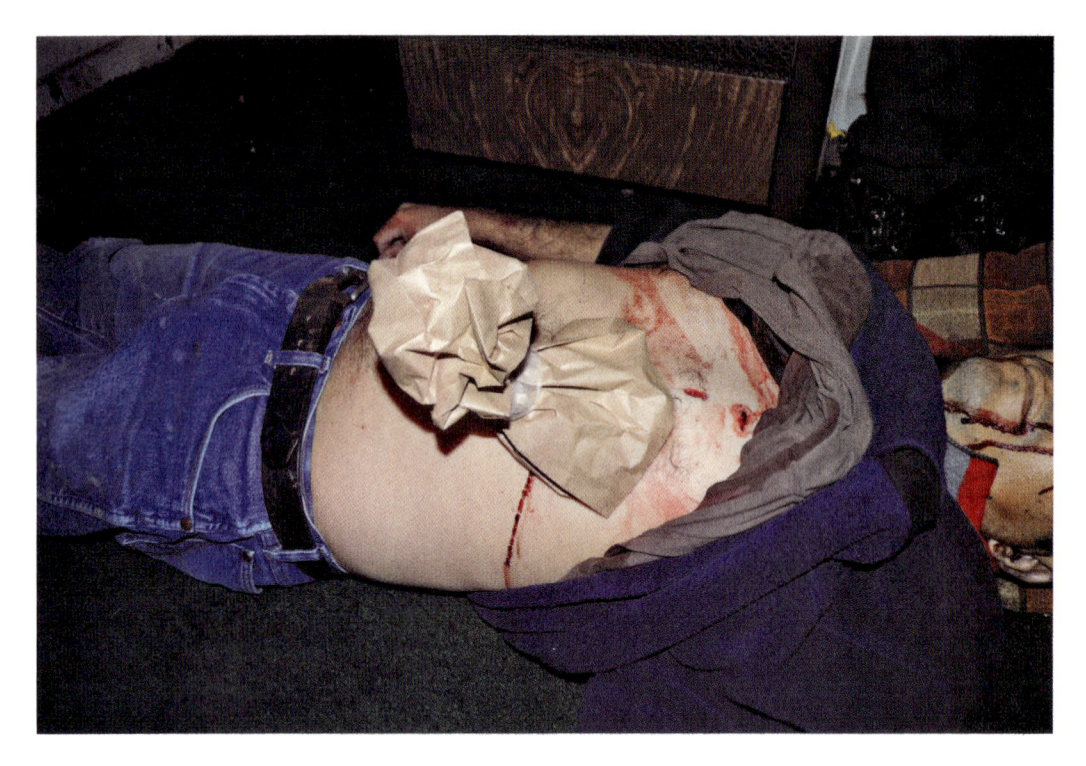

Figure 3.4 Fingerprints on the knife handle should be protected with a taped paper bag. The bag will protect them from loss or contamination by bodily fluid during transport.

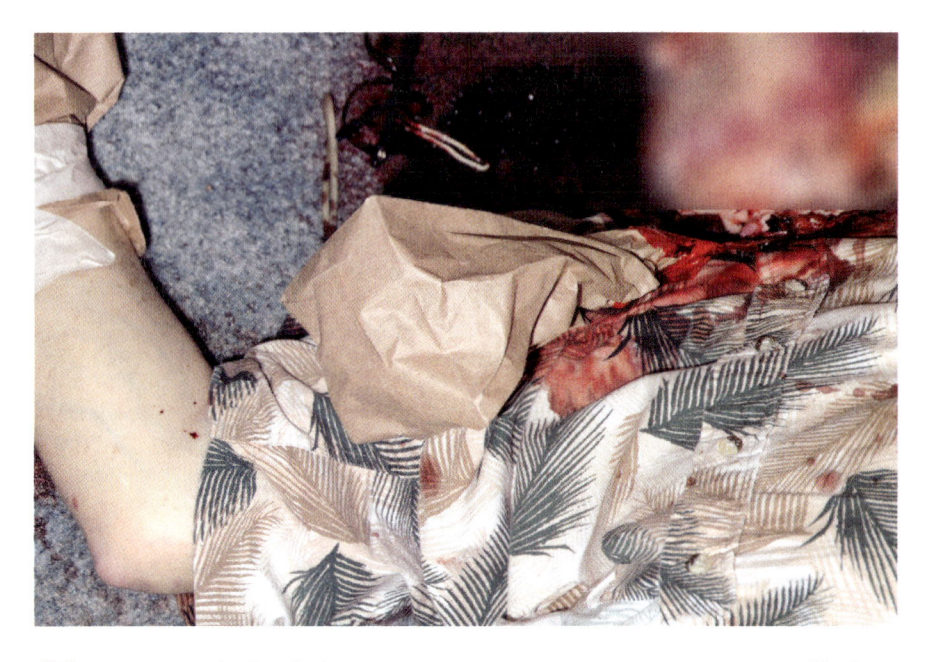

Figure 3.5 Preserving the knife during transport is important to protect possible fingerprints and/or DNA.

Preservation and Packaging 41

Figure 3.6 Once the duct tape is properly protected, it can be processed for fingerprints or touch DNA in the laboratory. The photo shows an example from a properly preserved segment of duct tape from a decedent.

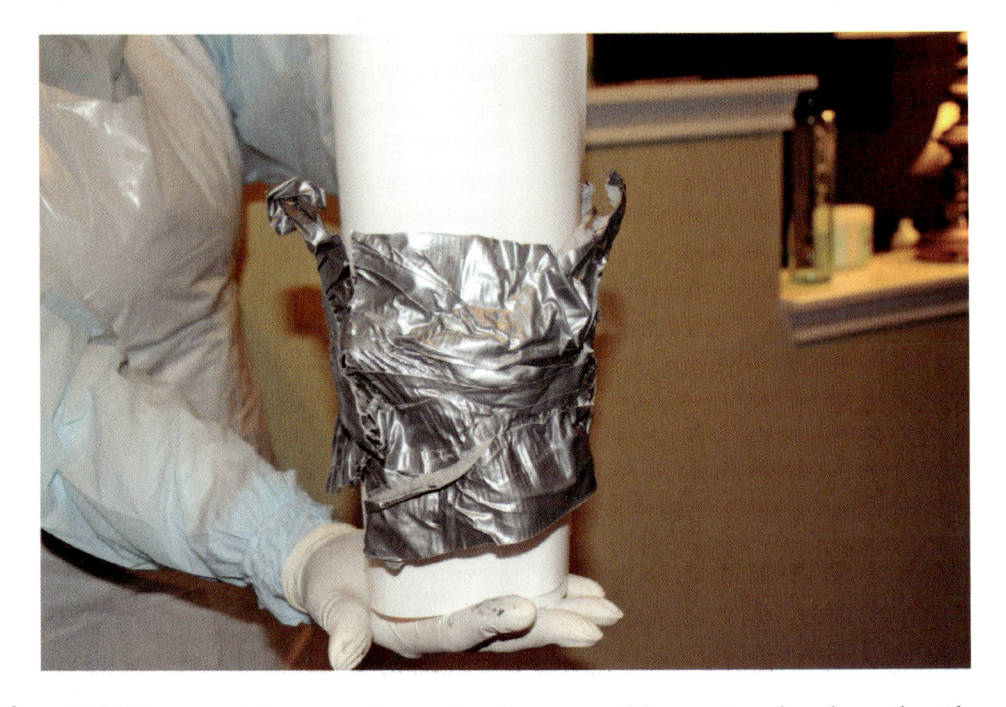

Figure 3.7 Duct tape is known to be an effective source of fingerprints, but the sticky side can also be a good source of touch DNA. Duct tape on the decedent may also be tested for fracture match to a roll of tape. Therefore, duct tape should be removed from the decedent, if possible, and placed on the shiny side of a piece of laboratory bench paper. The paper is then placed within a paper bag to ensure that it is protected from damage to fingerprints or DNA that may be on the tape.

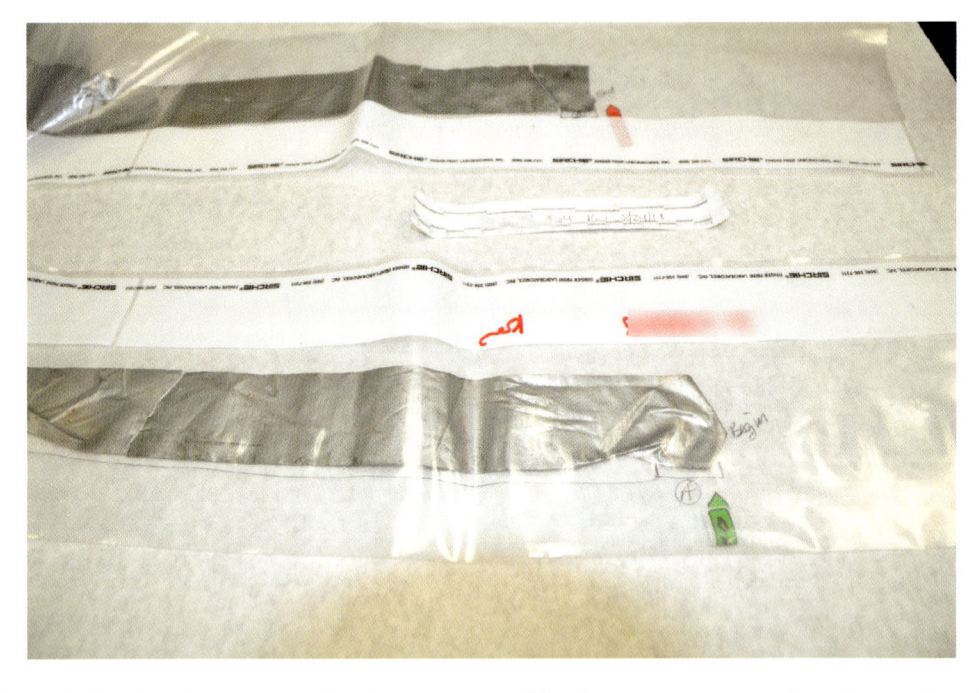

Figure 3.8 After the duct tape has been processed for fingerprints, it can be submitted to the lab for DNA analysis.

Preservation and Packaging

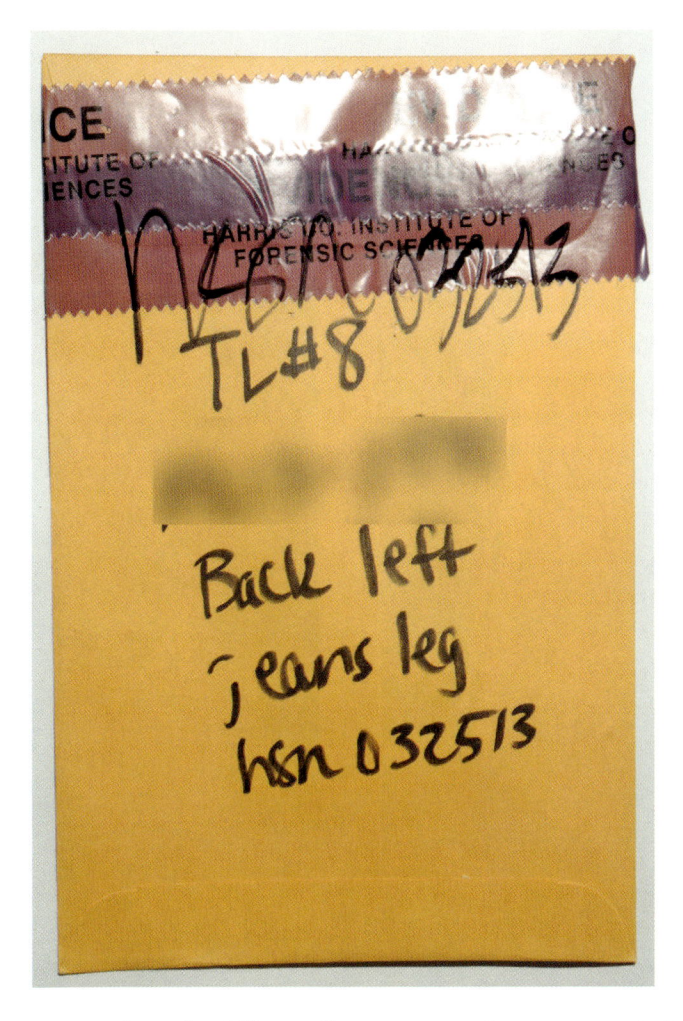

Figure 3.9 The paper envelope should have all important information, including case number, item number, date, and collecting analyst's initials.

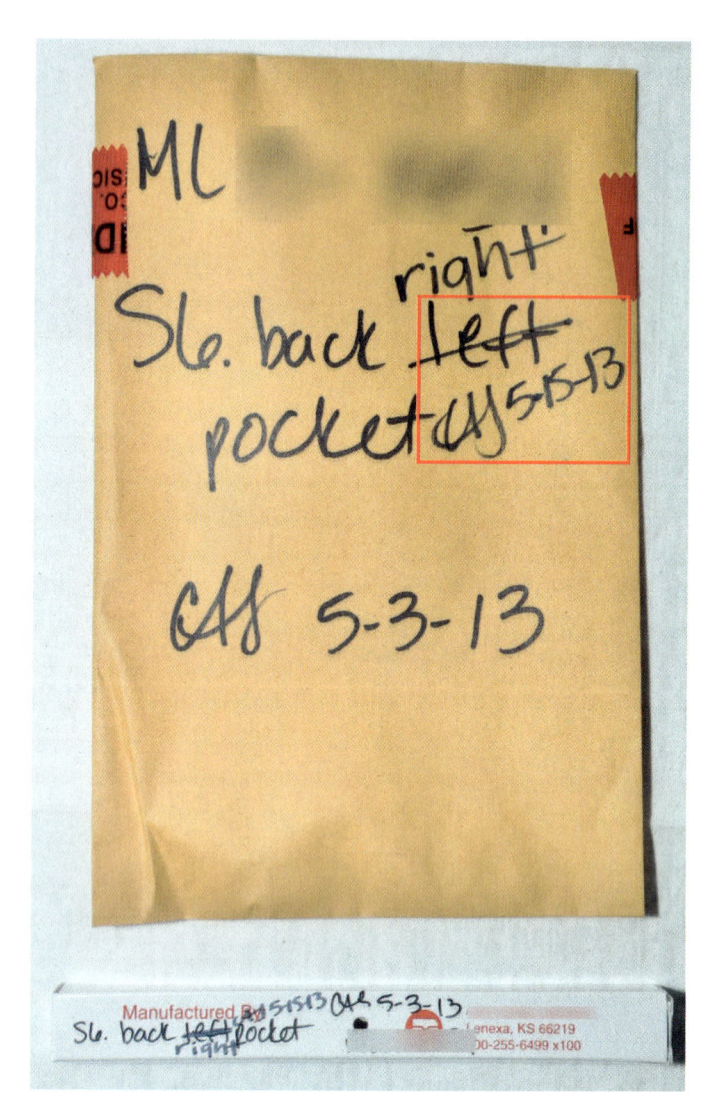

Figure 3.10 Errors must be corrected. The error must be lined out with a single line, initialed, and dated. The correct information should be written near the crossed-out information. The correction should be photographed to document it.

Preservation and Packaging 45

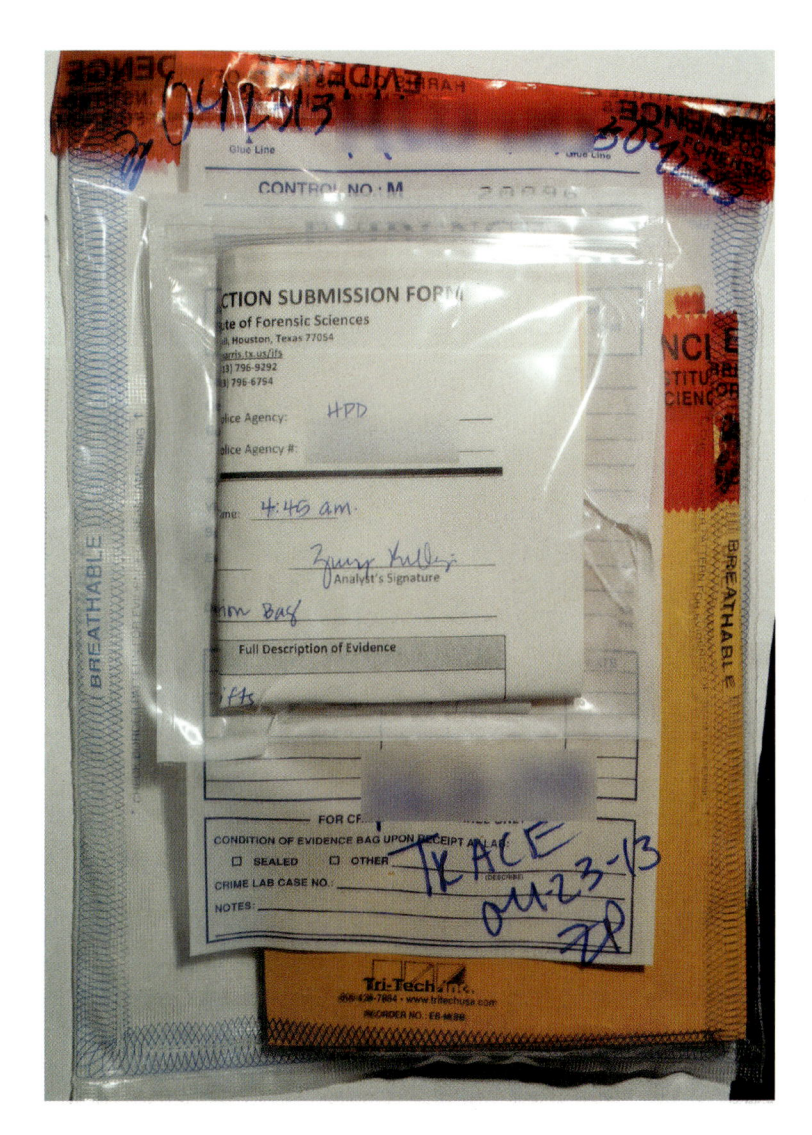

Figure 3.11 All collected evidence is placed in a plastic, breathable pouch for transport to the agency. The case number, collecting analyst, and the date are written on the pouch.

Figure 3.12 If the evidence item is larger than a paper envelope, it should be placed in a paper bag. If the item is wet, it must be dried prior to placing the item inside the paper bag. If transport is required while the item is wet, it can be placed in plastic for a short time and dried prior to final packaging.

Preservation and Packaging

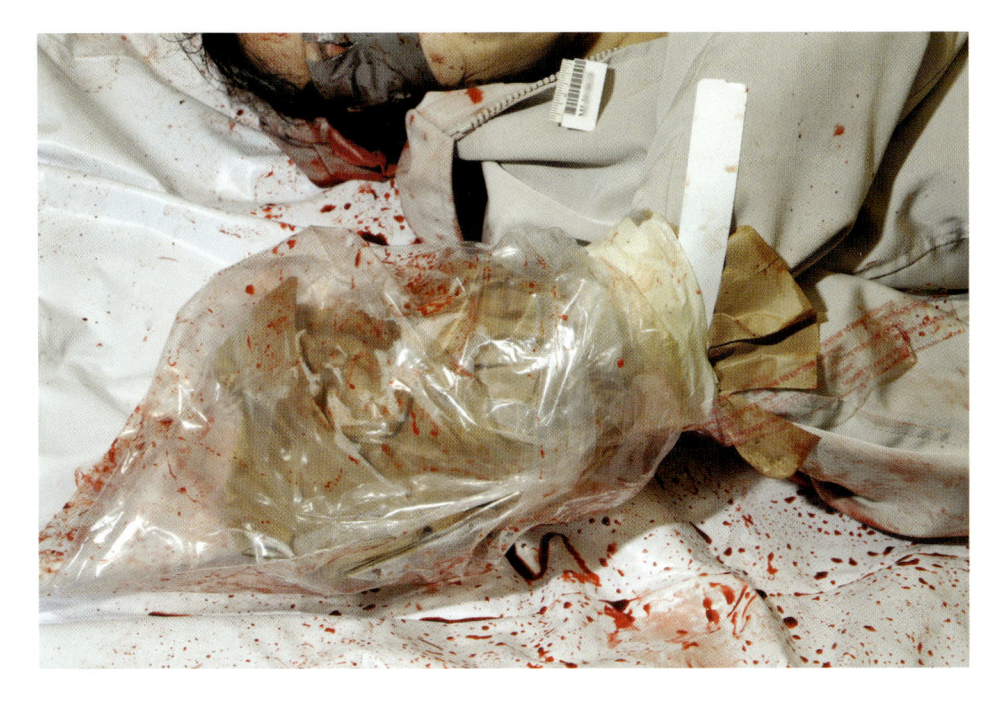

Figure 3.13 Paper bags are an effective way to protect hands and foreign objects. However, when hands are extremely bloody or decomposed, a plastic bag may need to be placed over the paper bag during transport to keep the paper bag from ripping or tearing and exposing the hand.

Figure 3.14 Unexpectedly, pockets have proven to be very good sources of foreign DNA. In instances where it is necessary to protect a pocket, it can be cut out with a scalpel and packaged in a paper envelope. DNA from the pocket can then be recovered and tested in the DNA laboratory.

Figure 3.15 If there are bloodstains found in the pocket of a decedent, it is best to cut the pockets out to preserve the blood patterns for downstream DNA analysis.

Figure 3.16 Dry areas of clothing may be cut and removed to preserve stains or possible areas of touch DNA.

Preservation and Packaging

Figure 3.17 The decedent's clothing may be completely saturated if there is decomposition. If a dry area can be found, it should be swabbed or protected for DNA analysis. In this example, the lower pant legs were the only dry areas on the clothing of the decedent. The pant legs were cut off and packaged in a brown bag to preserve any blood patterns or touch DNA.

Figure 3.18 Hands of homicide victims should be "bagged." Bagging will protect gunshot residue as well as foreign DNA that may be found under the fingernails.

Figure 3.19 Even if the hands of a decedent are bound, the hands and bindings should be bagged to preserve gunshot residue as well as foreign DNA.

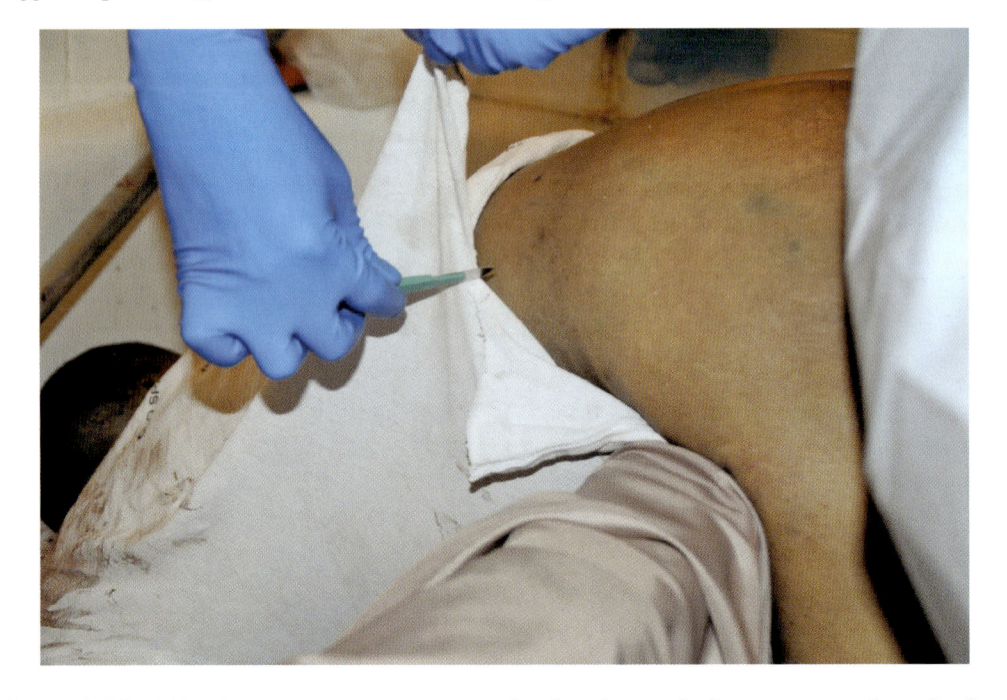

Figure 3.20 If blood patterns are present on the decedent's clothing, protect them for later examination. For example, this shoeprint might be compared later to prints from a suspect's shoe. Alternatively, the clothing can be cut off the decedent and protected in an envelope or paper bag. It is important not to cut clothing near a stab or gunshot wound. Those clothing penetrations must be preserved for the pathologist.

Preservation and Packaging 51

Figure 3.21 Once an item is cut from the decedent for preservation, it should be documented by a photo and placed in a paper bag for storage.

Body Swabbings

4

Swabbings from the body are crucial when trying to collect foreign touch DNA, saliva, blood, or semen stains. The technique for swabbing on the body is similar to swabbing material evidence items, except that the swabbing requires less force. Swab with medium pressure to remove only the top layer of stain or cells while minimizing the removal of the decedent's cells.

When at the crime scene, swabbings of the body are limited to external surfaces. If deemed necessary, the pathologist can collect evidence from internal orifices in the morgue, in conjunction with examination of those areas as part of the sexual assault examination.

When approaching the decedent at the crime scene, imagine how and where the assailant might have touched the decedent—during a struggle or after death. Focus collections on all of the areas, such as armpits, ankles, belts, and waistbands, that the assailant would have touched. Imagine the events that led to the placement of the body in its current location. For example, if it appears the body was dragged by the ankles, you will want to focus your collection efforts at the ankles. If you find an area on the body devoid of blood, that would be a good place to swab.

If the decedent and suspect had opportunities for contact between them prior to the death (e.g., they are known to have spent time together or they lived together), extraneous transfer of DNA may occur that is not related to the investigation. However, we have found that collection can be worthwhile even in these situations, as DNA foreign to the suspect and the decedent may be recovered. In one study (Graham and Rutty, 2008), DNA transfer between roommates was observed only 23% of the time, and alleles foreign to the roommates were also observed. In our view, collection is worthwhile when the decedent is bound, when there is evidence that the decedent was moved, or when there is evidence that the decedent had contact with an assailant.

We have had success in obtaining foreign alleles by swabbing the decedent's knuckles and fingertips (Figures 4.1 to 4.3), neck (Figures 4.4 to 4.6), face (Figures 4.7 and 4.8), breasts (Figure 4.9), ankles (Figures 4.10 and 4.11), arms (Figure 4.12), belly button (Figure 4.13), unique staining (Figures 4.14 and 4.15), and the penis (Figure 4.16), if any of these areas are exposed. Body jewelry should also be swabbed if present on the body (Figure 4.17). Bruises in exposed areas (Figures 4.18 and 4.19), voided areas around wounds (Figure 4.20), or dismembered areas (Figure 4.21) should be swabbed regardless of location. Also, swab voids in bloodstained areas (Figures 4.22 and 4.23). Voids areas are not contaminated with the decedent's blood and may be where the assailant touched the decedent (hence protecting that area from blood deposition/smearing). However, be cognizant that some blood on a decedent may in fact not have originated from the decedent—this may be true even when the body is very bloody. This we discuss in Chapter 4.

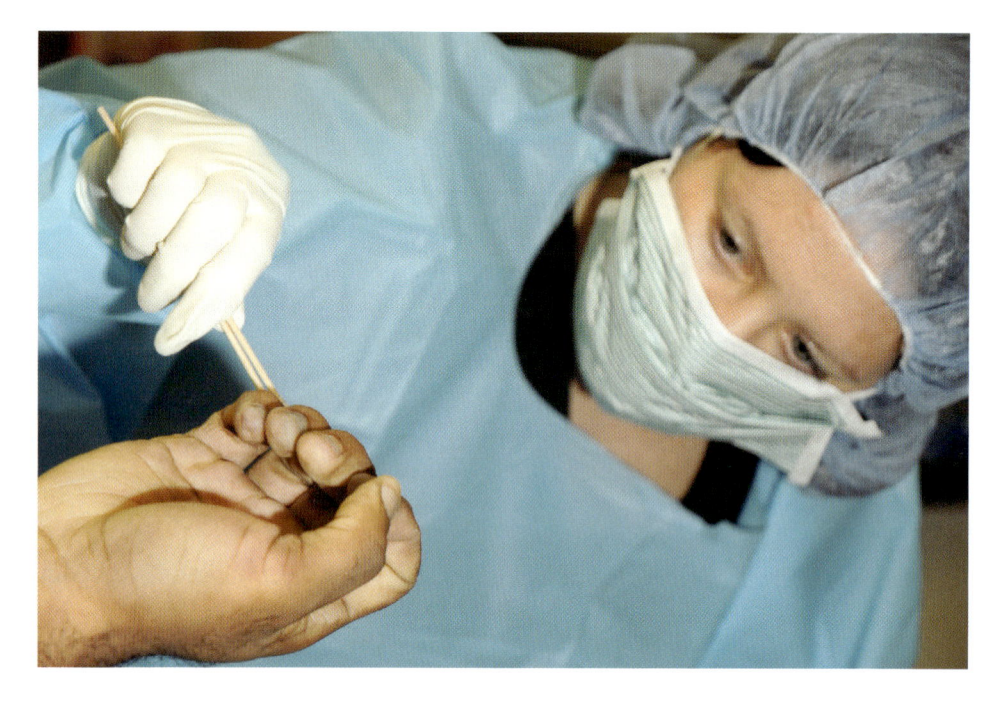

Figure 4.1 Knuckles are a key area to swab for trace collection for foreign DNA. If there is blood present on the knuckles, it is best to swab the unstained areas to avoid decedent blood. Swab the knuckles with medium pressure in an effort to collect transferred cells from the assailant while minimizing collection of decedent's cells.

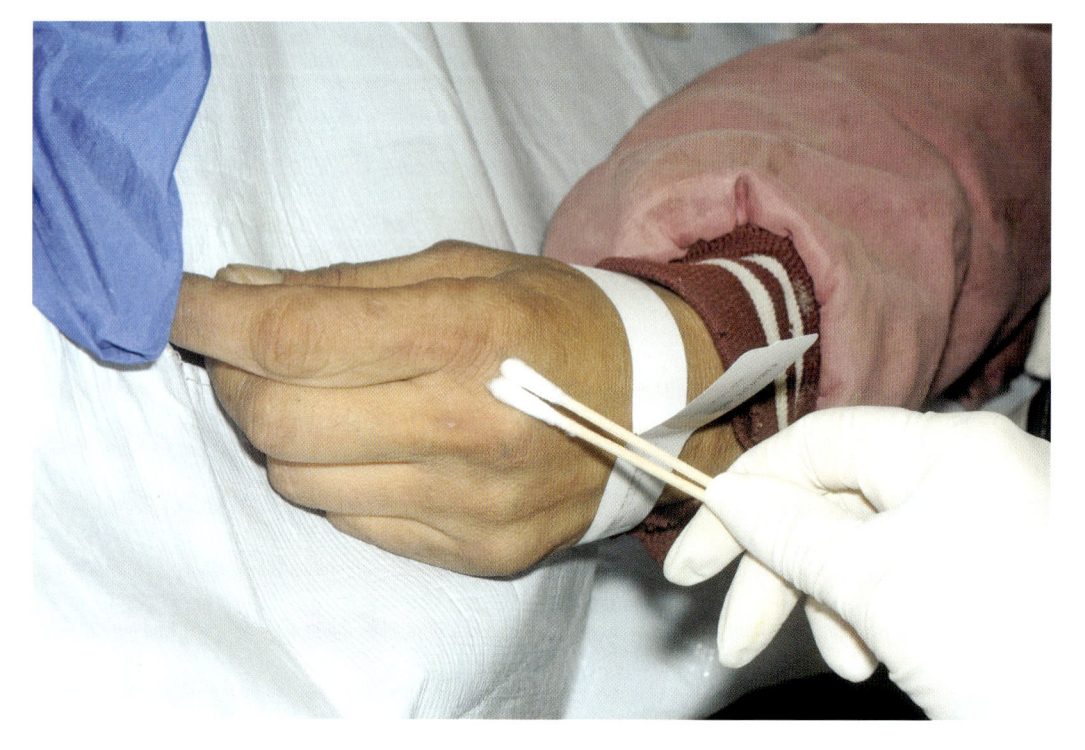

Figure 4.2 Swabbing of knuckles and fingertips are critical areas to swab when looking for foreign DNA.

Body Swabbings 55

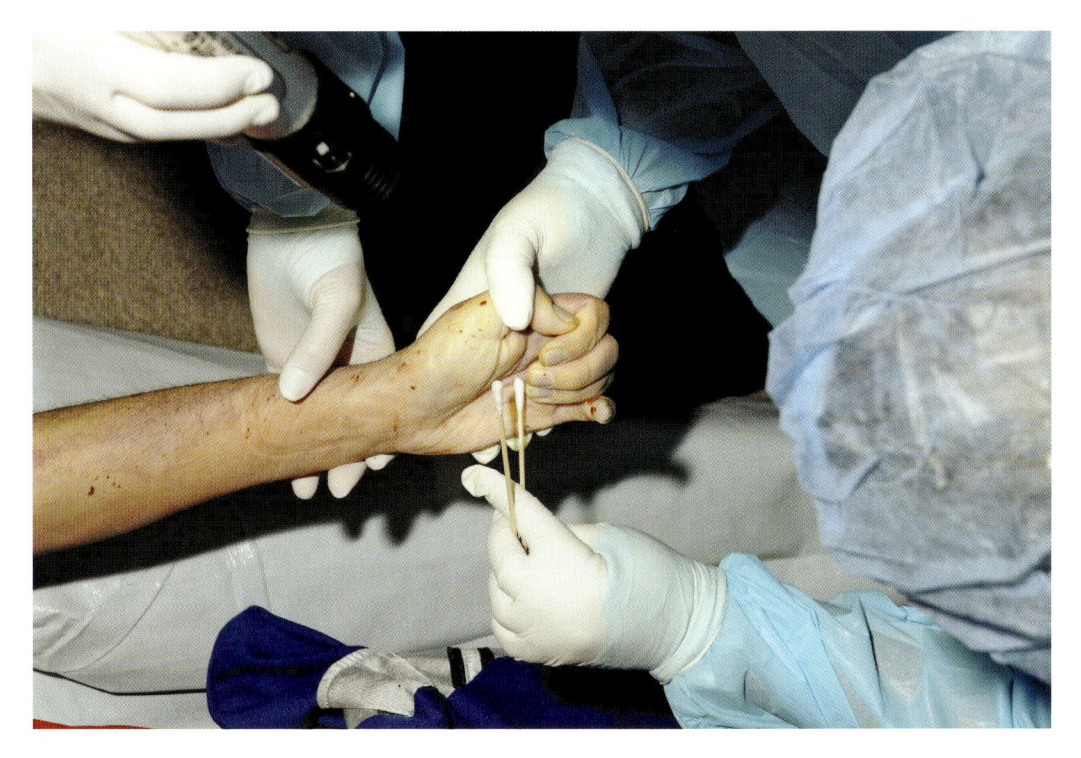

Figure 4.3 The fingertips are also a good area to swab for foreign DNA. If the decedent fought, he or she may have collected foreign DNA or fibers from his or her attacker. Fingernails are clipped by the pathologist, but the fingertips can be swabbed prior to transport of the decedent to the morgue.

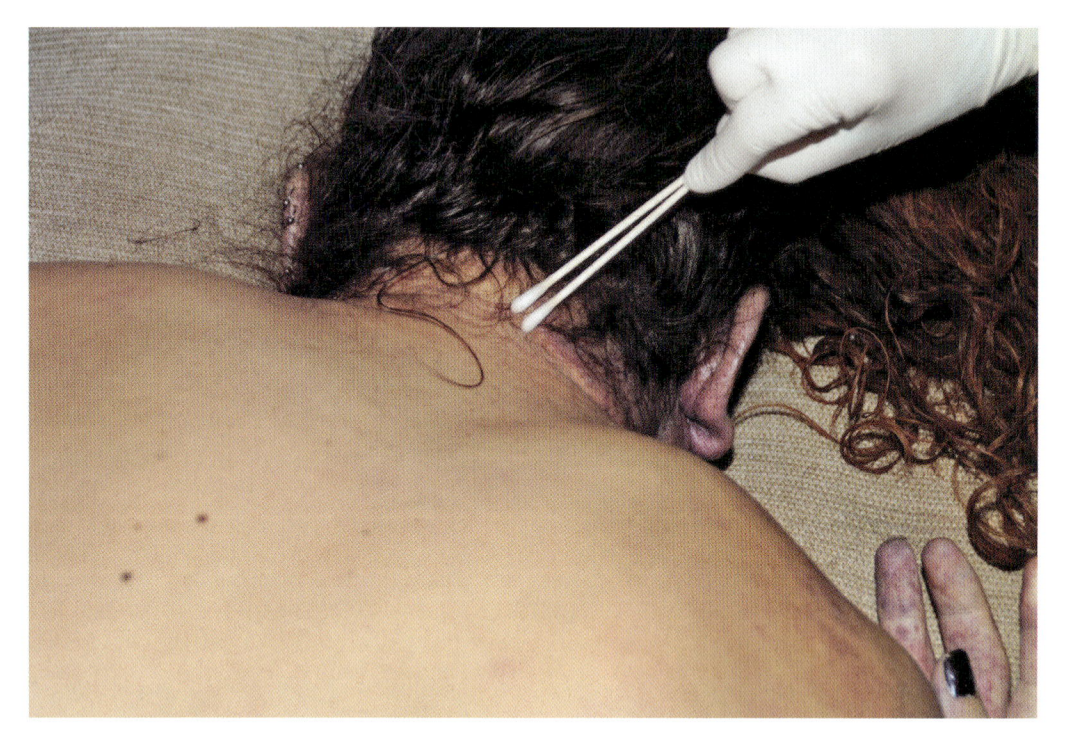

Figure 4.4 The neck is also a good place to find foreign DNA if the decedent were choked or strangled. Often the friction during strangulation will leave foreign DNA that can be recovered. Swabbing the neck with medium pressure will allow the surface cells to be collected with a minimal number of the decedent's cells.

Body Swabbings 57

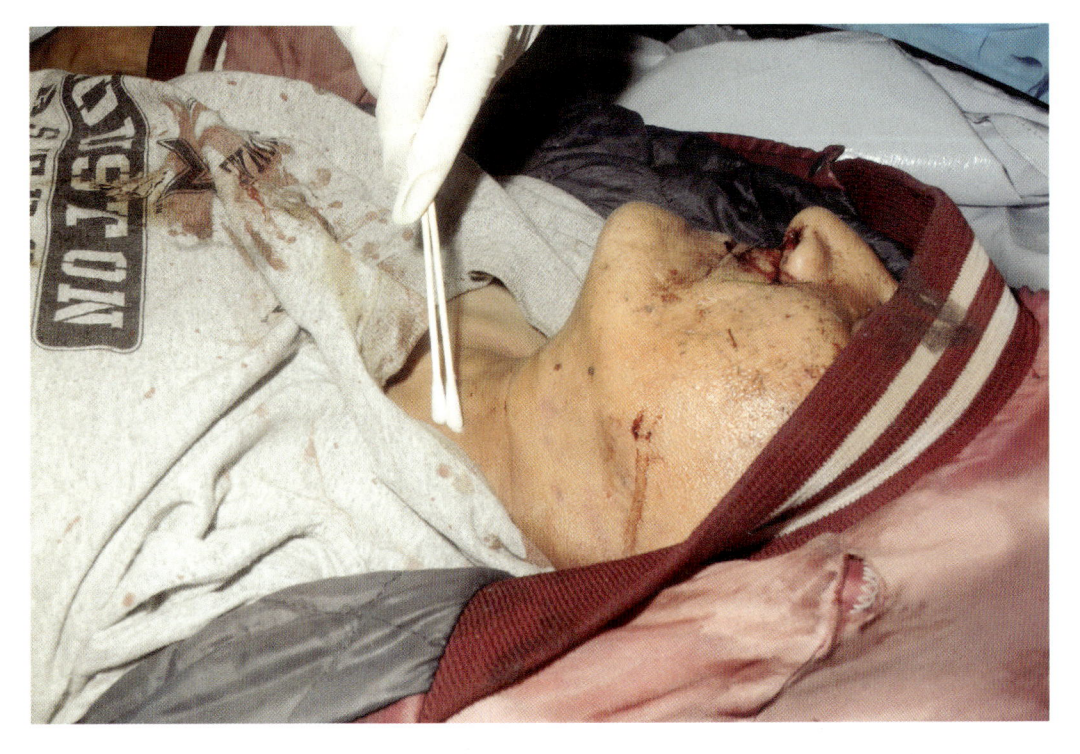

Figure 4.5 The neck may contain foreign DNA that can be used to determine the identity of a suspect in the case.

58　　　　　　　　　　　　Forensic DNA Collection at Death Scenes: A Pictorial Guide

Figure 4.6 Neck swabbings must be performed with a medium pressure to ensure minimal decedent DNA is collected.

Body Swabbings 59

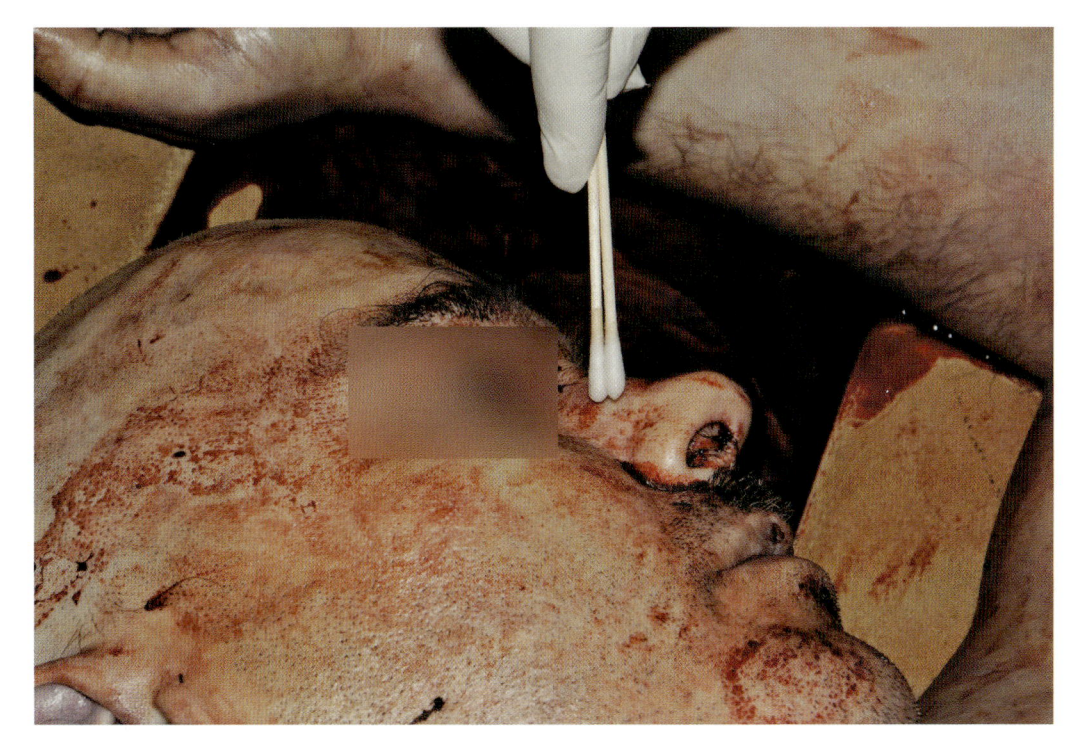

Figure 4.7 The face of the decedent may have been touched or punched prior to death. If there is an unstained area, it may be a good place to swab for foreign DNA. Avoid any blood that is present.

Figure 4.8 The lips of the decedent can yield foreign DNA. If the decedent is found nude or if there are other indications of sexual relations prior to the death, the lips are swabbed for foreign DNA.

Body Swabbings 61

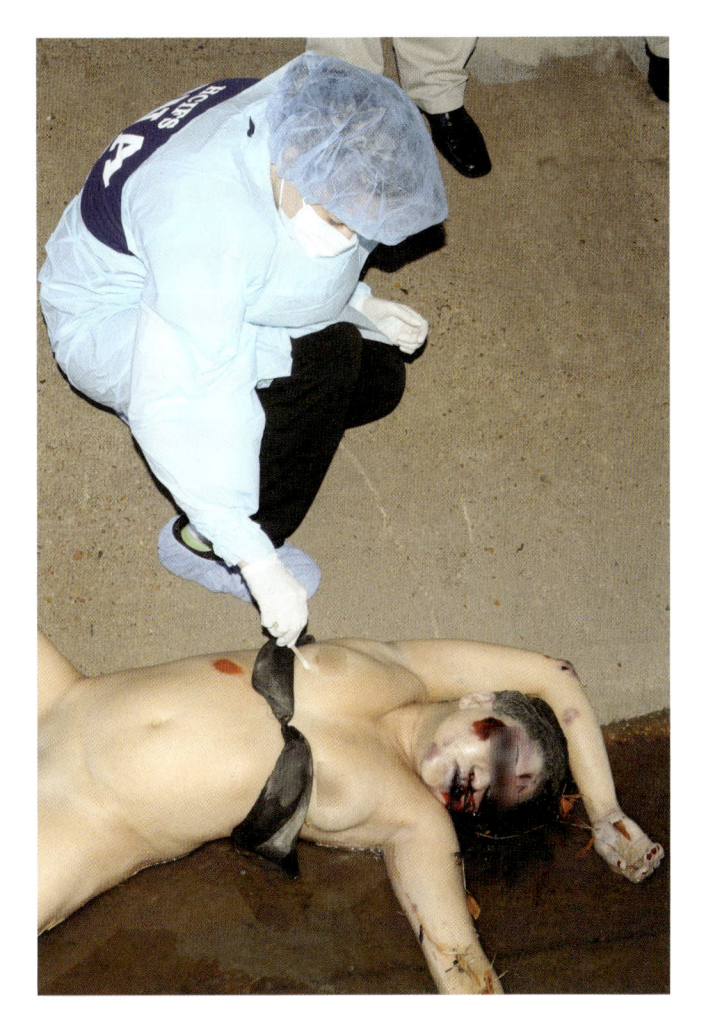

Figure 4.9 If the decedent is a naked female, swab the breast area, as this area may have been touched during a sexual act. Swab with medium pressure to optimize collection of foreign cells while minimizing collection of cells from the decedent.

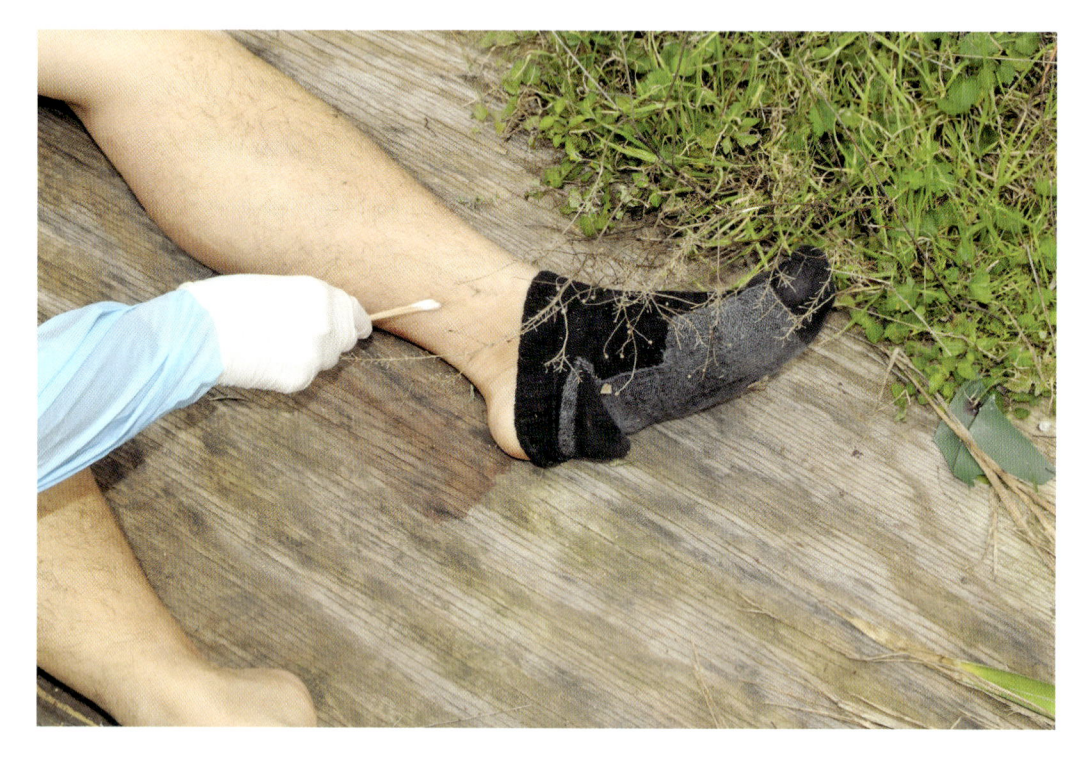

Figure 4.10 The ankles of a decedent are a good place to collect samples for foreign DNA. Assailants will drag the body by the ankles to move it. This is likely when blood is present on the upper but not the lower part of the body.

Body Swabbings

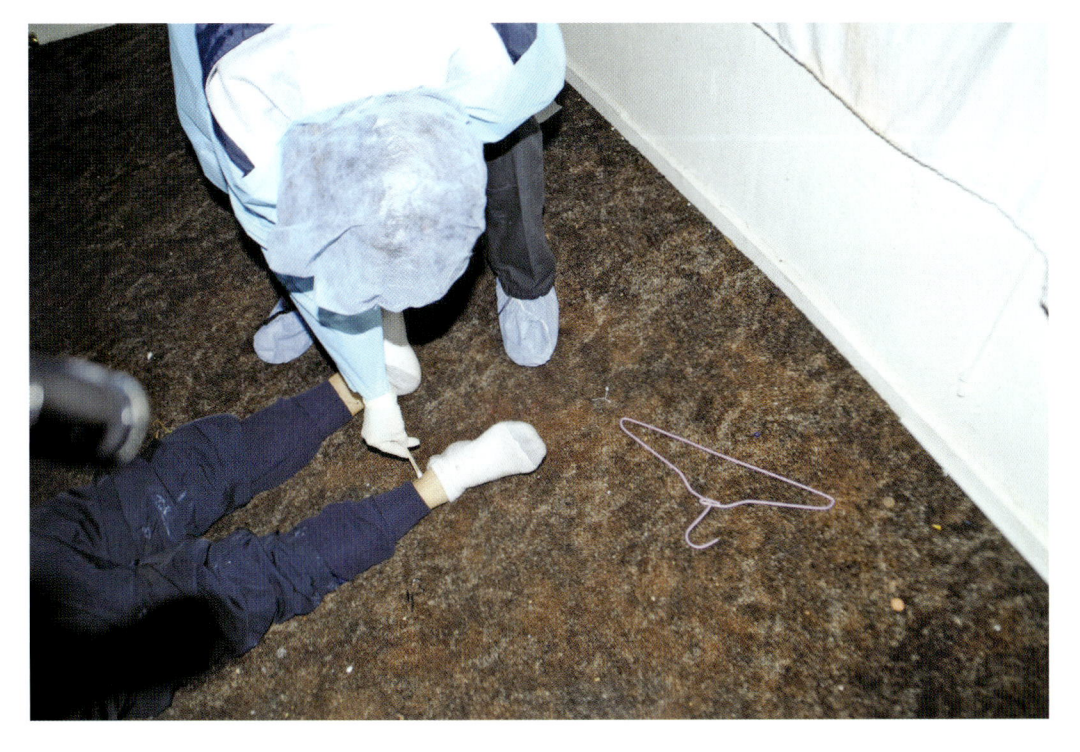

Figure 4.11 If socks are on the ankles, swab both the ankles and the top of the socks. Assailants may grab the socks while dragging the body.

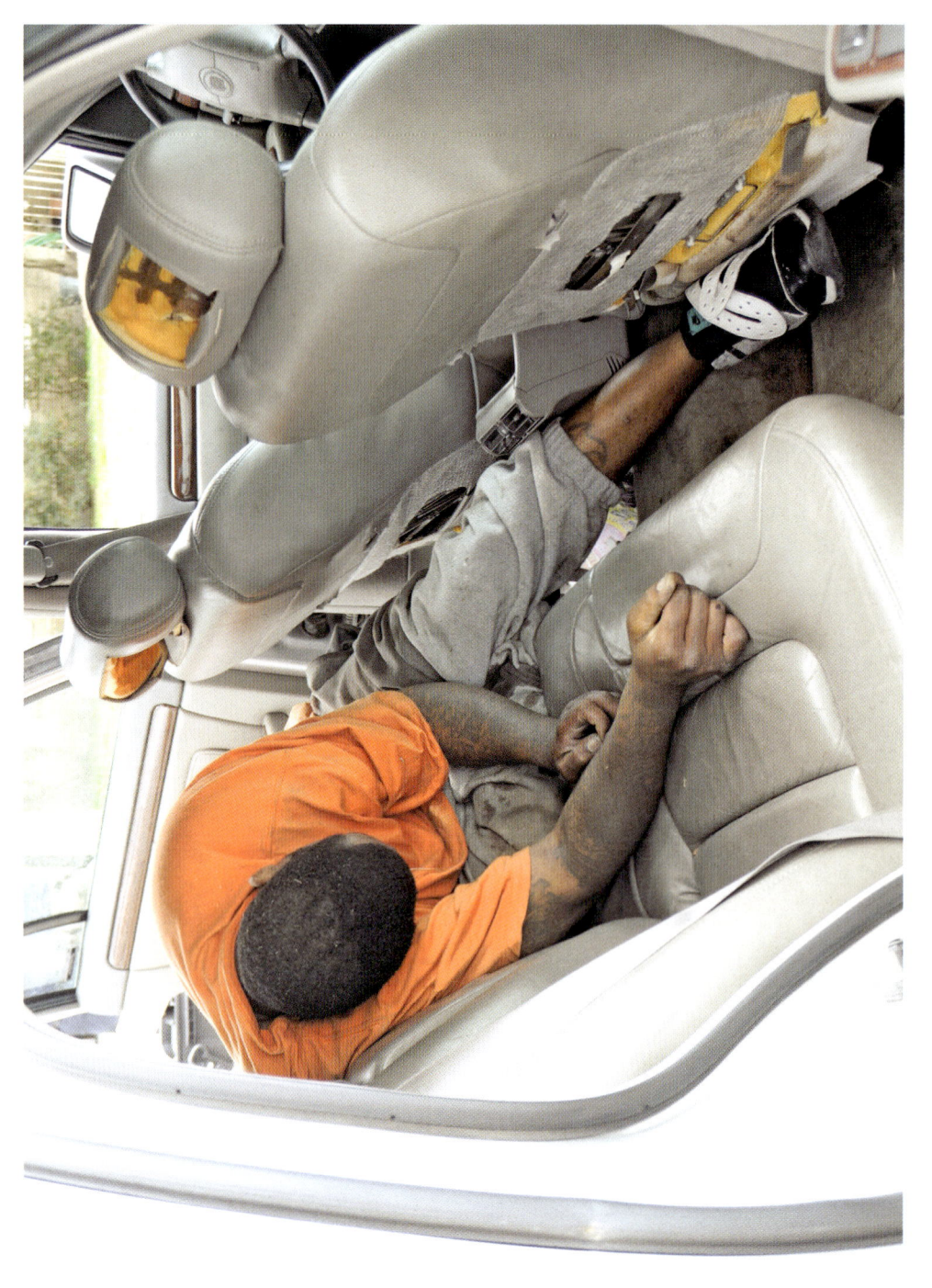

64 Forensic DNA Collection at Death Scenes: A Pictorial Guide

Figure 4.12 Arms and wrists are likely places of contact for an assailant dragging the body or while attacking the decedent.

Body Swabbings 65

Figure 4.13 Body jewelry can be a source of foreign DNA. Depending on where it is on the body, it can yield foreign DNA. Belly button rings are a good place to swab, especially if the decedent is naked. If there are indications of sexual relations prior to the death, the belly button ring may have trapped the assailant's cells.

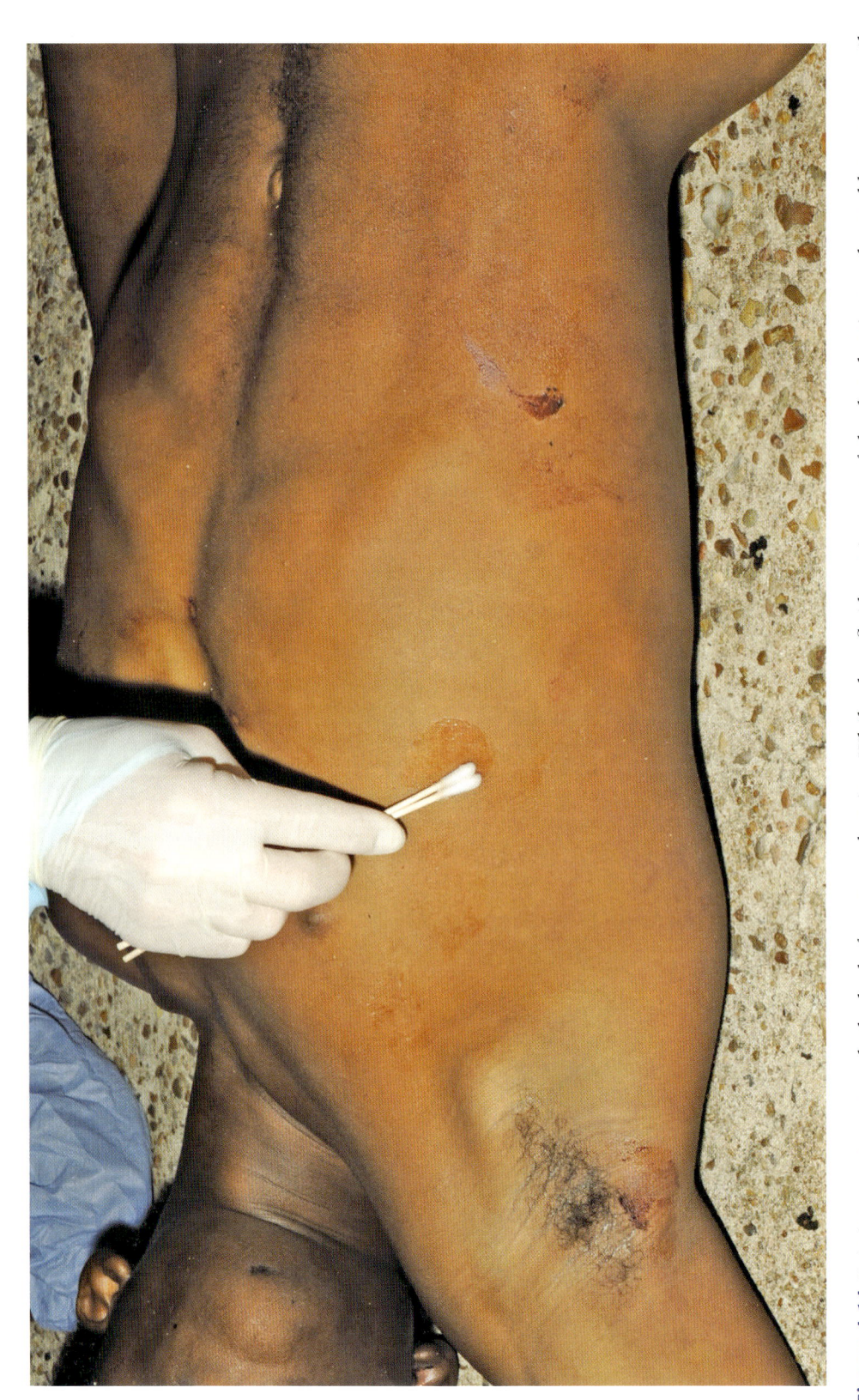

Figure 4.14 It is important to search the body for unusual stains. While clear fluid coming out of the decedent's mouth would not necessarily be important, a clear stain on the torso not near a wound might seem out of place. It could be semen, saliva, or sweat and should be collected.

Body Swabbings 67

Figure 4.15 If at the crime scene the decedent has a condom on the body, swab where it is touching the body and package the condom for testing in the DNA laboratory.

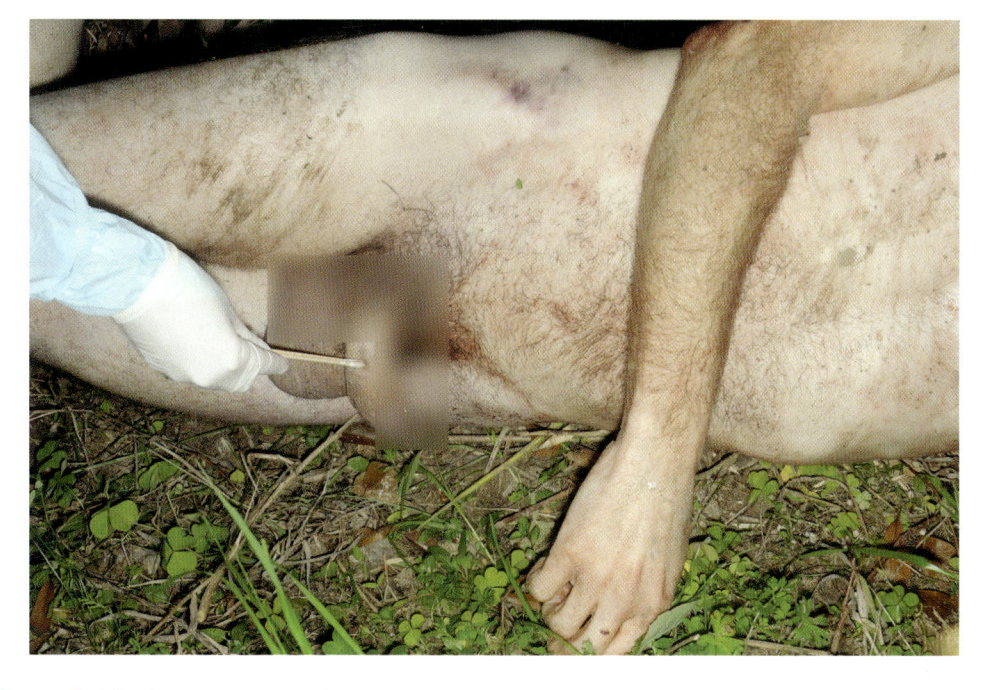

Figure 4.16 The penis is an obvious place to swab a naked male decedent for foreign DNA from a sex act. This area is swabbed because it cannot be protected well during transport.

Figure 4.17 Rings are good places to find touch DNA. Perhaps the decedent punched the assailant or vigorously fought with him or her. The assailant's DNA may be in the crevices of the ring.

Body Swabbings

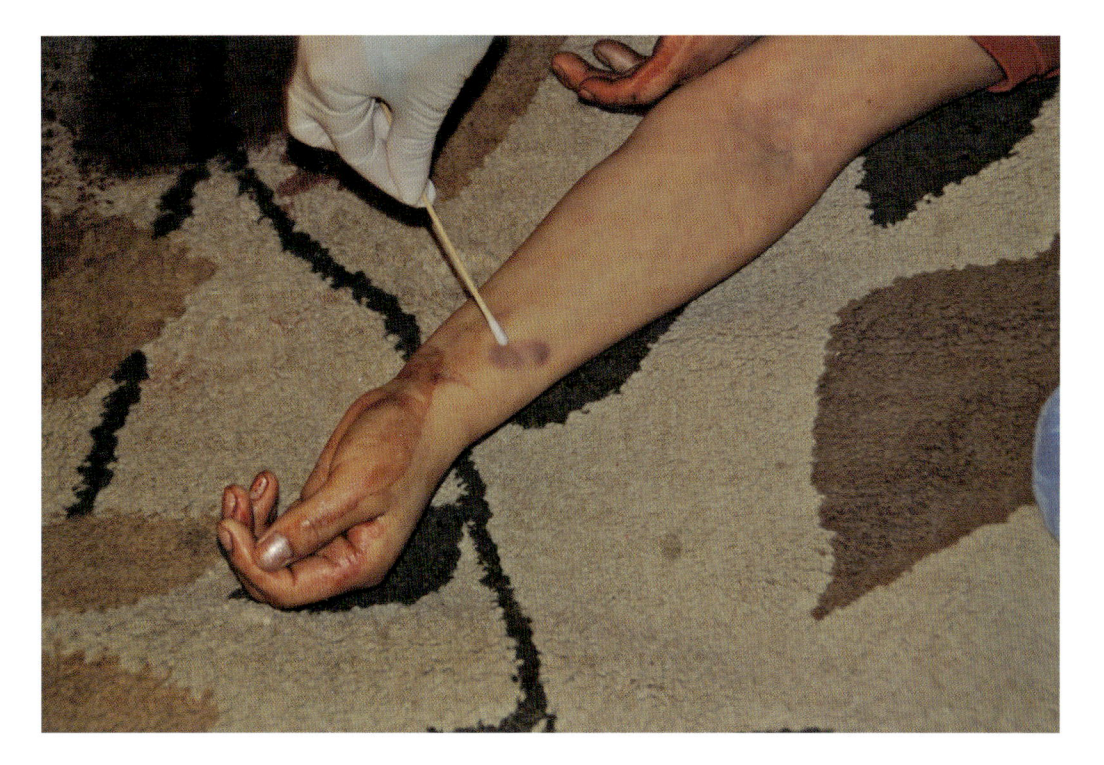

Figure 4.18 A bruise should always be swabbed, as it may indicate an area where the assailant came in contact with the decedent.

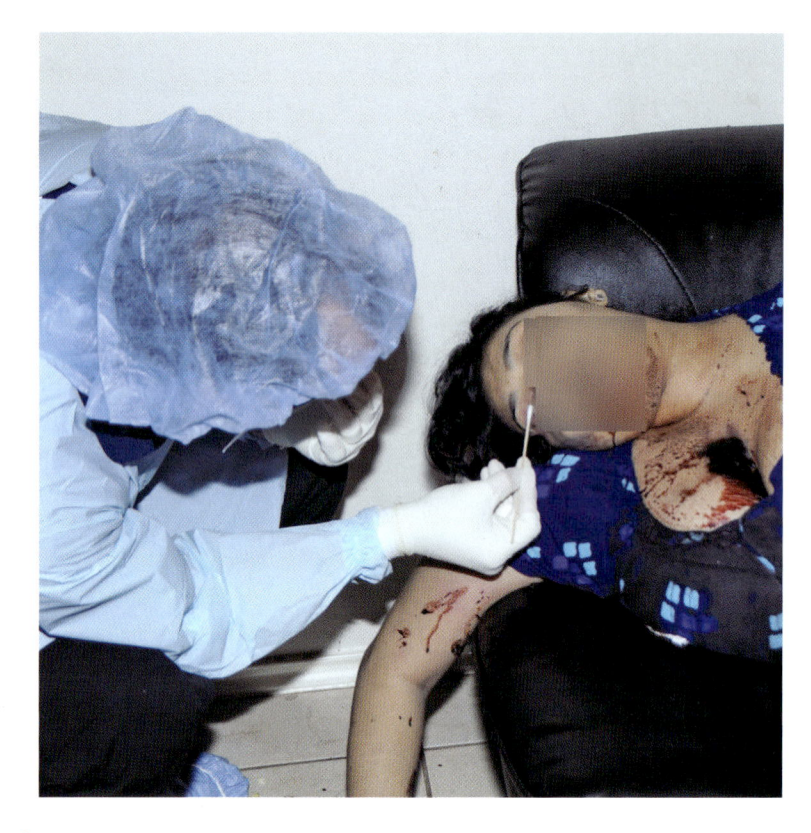

Figure 4.19 Bruises are likely areas to find foreign DNA and should be swabbed.

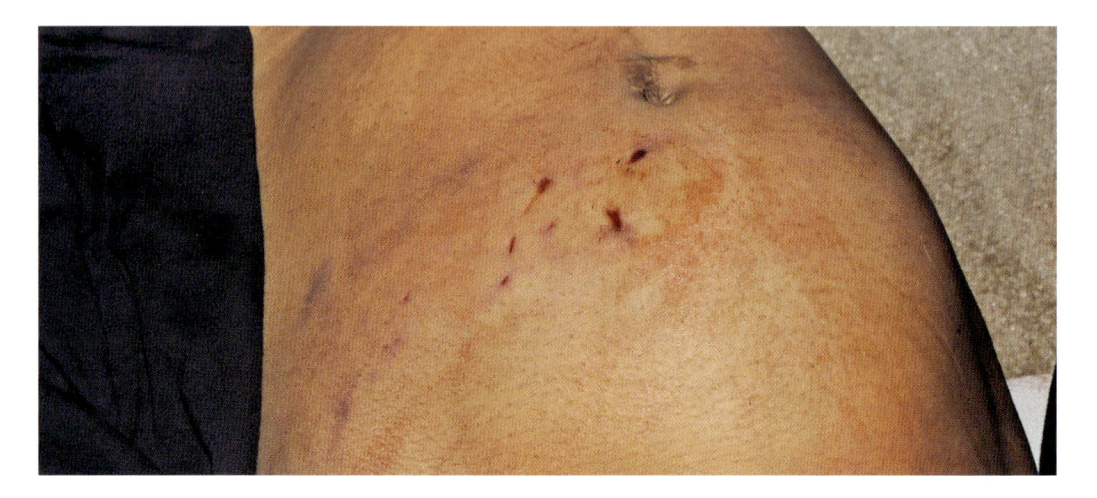

Figure 4.20 If there is apparent trauma to the face, foreign DNA may be found in the area of the trauma. Swab the area while avoiding any visible blood on the decedent. The photo shows stab wounds; the assailant may have touched the decedent with his or her hand while stabbing.

Body Swabbings 71

Figure 4.21 When a body has been dismembered, it is important to think about the areas that would have been touched by the assailant during the process. If the head was removed, the shoulder and neck area should be swabbed. Swab the shoulders if the arms have been removed.

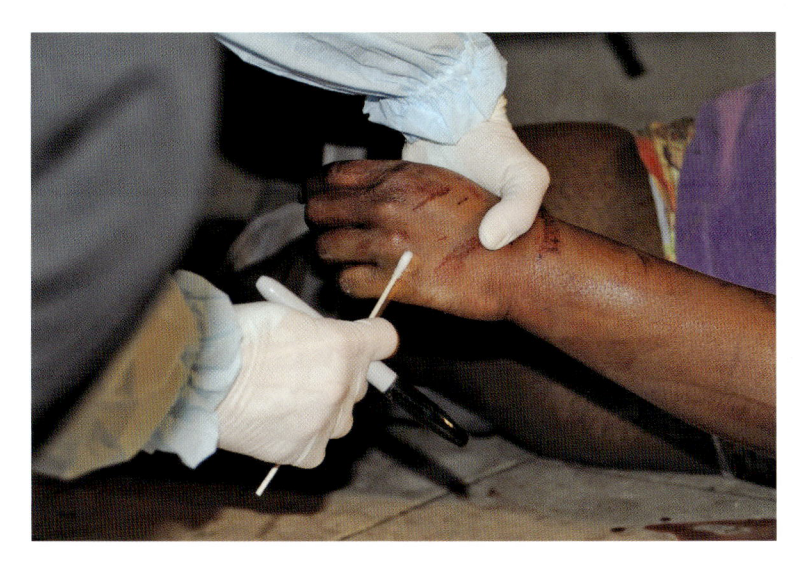

Figure 4.22 When swabbing key areas of interest on the body, it is best to swab in the areas that do not have visible blood or bodily fluids present, as these areas have a better chance of yielding foreign DNA without the decedent's DNA profile. It is also more likely that the assailant touched those areas. Perhaps the assailant touched the body there to avoid getting blood on his or her hands, or perhaps the assailant's hand blocked the blood and caused the void.

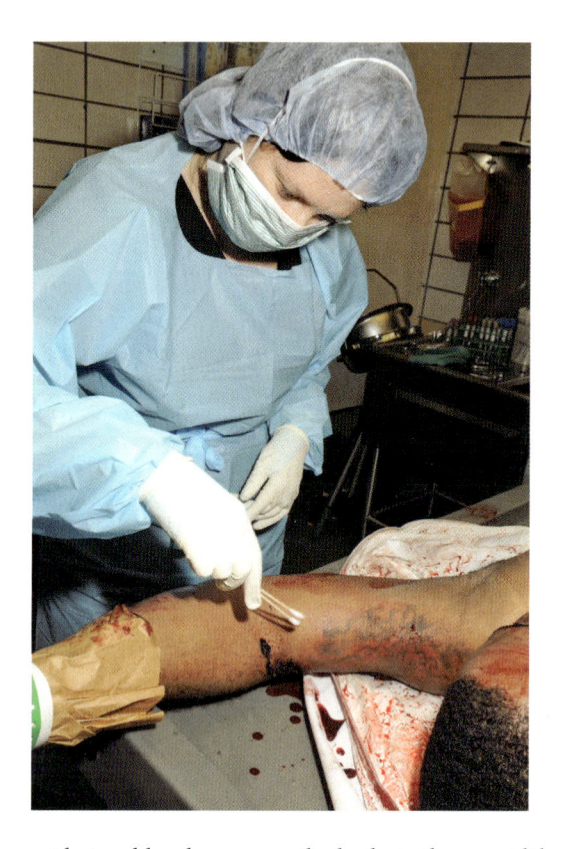

Figure 4.23 Swabbing voids in a bloody area on the body is the most likely place foreign DNA will be found.

Blood Patterns

5

Blood patterns can be very helpful in the investigation of homicides. Passive drops, transfer/contact patterns, swipe patterns, wipe patterns, and void patterns are examples of characteristic patterns to note (SWGSTAIN, 2009). Passive drops, also known as 90 degree blood drops, indicate the blood source was at a 90 degree angle from the surface of the body. Ninety degree blood drops are not likely to be the decedent's blood and should be collected. Transfer/contact patterns are also important blood patterns. These patterns appear when a bloody surface is transferred to another surface. This type of pattern may indicate an area where the assailant's DNA transferred to the decedent. Swipe patterns are similar to transfer patterns, but the transfer pattern is directional. Directionality may be seen as pattern feathering at the edge where movement ended. Wipe patterns are similar to swipe patterns, except that the wet blood isn't transferred; it is already present as another object moves through the stain (SWGSTAIN, 2009). As mentioned earlier, a void pattern, a blank area within a bloody area on the body, may indicate where an assailant held or touched the decedent (Figures 5.1 to 5.14).

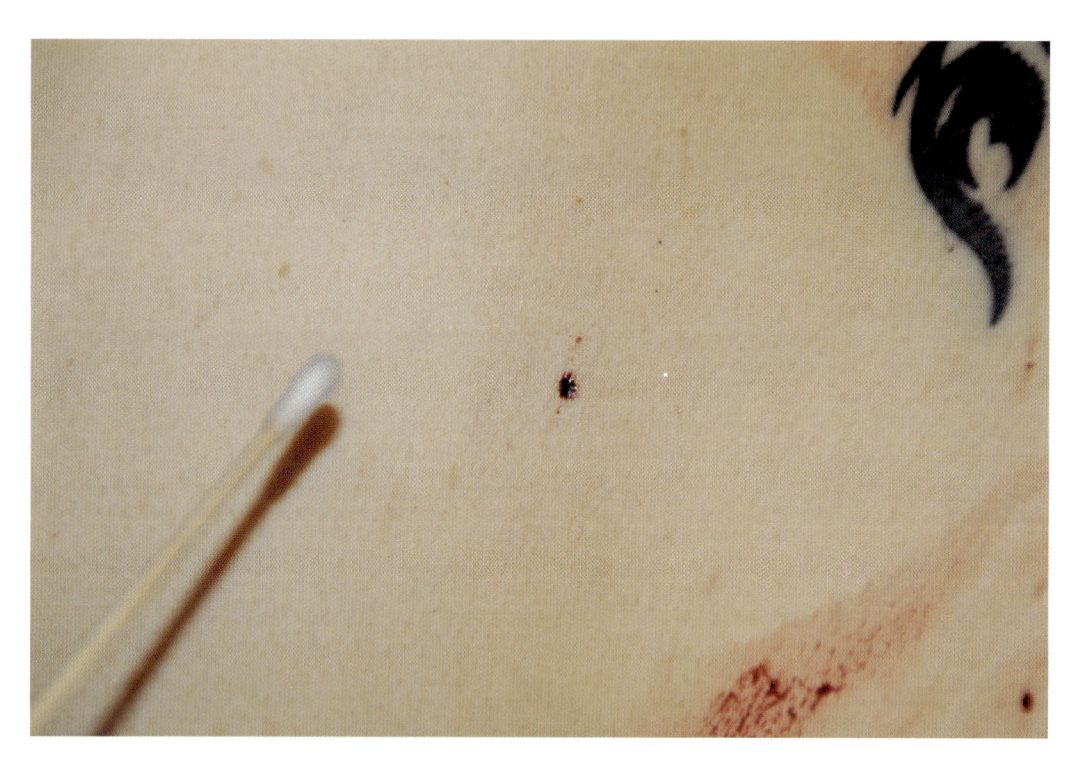

Figure 5.1 Ninety degree blood drops indicate the blood source was perpendicular to the decedent. Perhaps the assailant bled while standing over the horizontal decedent. A 90 degree blood drop is most likely not the decedent's blood. Ninety degree blood drops must be collected when they are found.

Blood Patterns

Figure 5.2 Ninety degree blood drops on shoes can be informative. A decedent could leave a perpendicular blood drop on his or her shoe, but the drop should be collected and tested in the event it yields a foreign DNA profile.

Figure 5.3 Blood drop patterns on shoes may end up being deposited from a suspect at the scene. All unique patterns should be swabbed to find out if any foreign DNA is present.

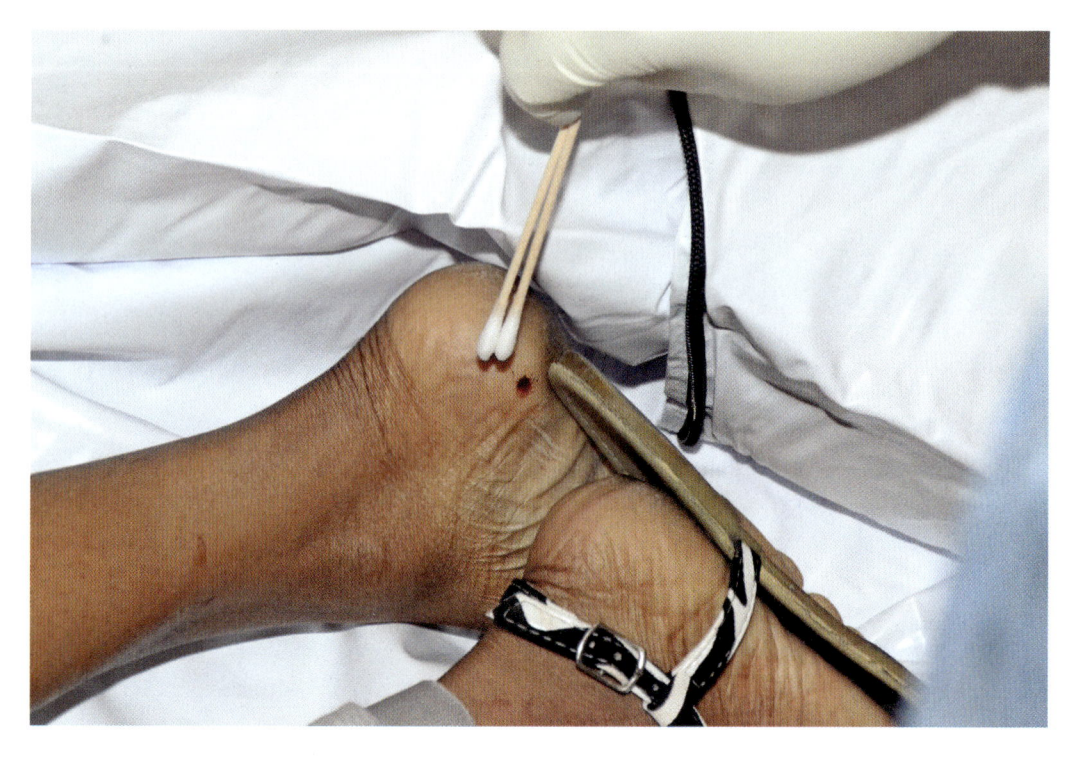

Figure 5.4 A 90 degree blood drop on the side of the foot is very likely to have come from someone other than the decedent.

Blood Patterns

Figure 5.5 The location and pattern of blood on the body can be very telling. If there is a smear of blood on the bottom of the foot, that is an unusual location for someone to touch himself or herself. It seems more likely that someone else transferred the bloodstain to this area, or maybe even hurt himself or herself and bled on the body. This area should be swabbed and collected.

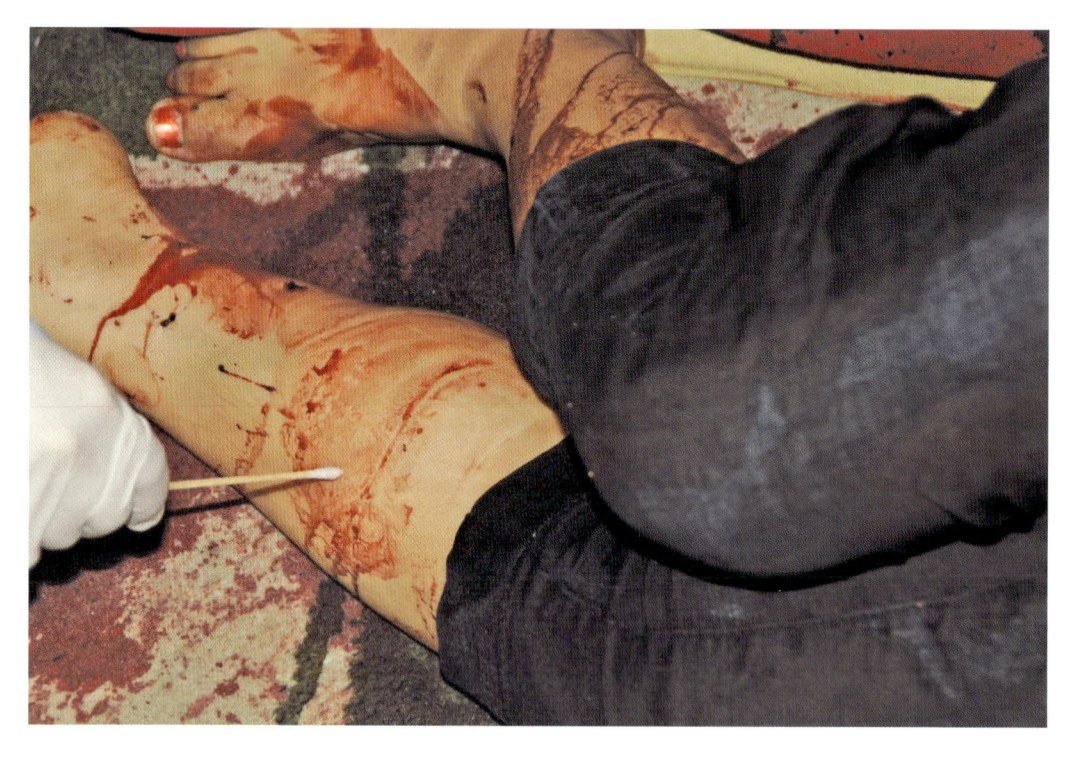

Figure 5.6 Blood swipes are deposited by someone smearing blood to another area by transferring the blood there. These swipes can be from the decedent, or they can be caused by a wound obtained during the murder that caused the suspect to bleed.

Blood Patterns

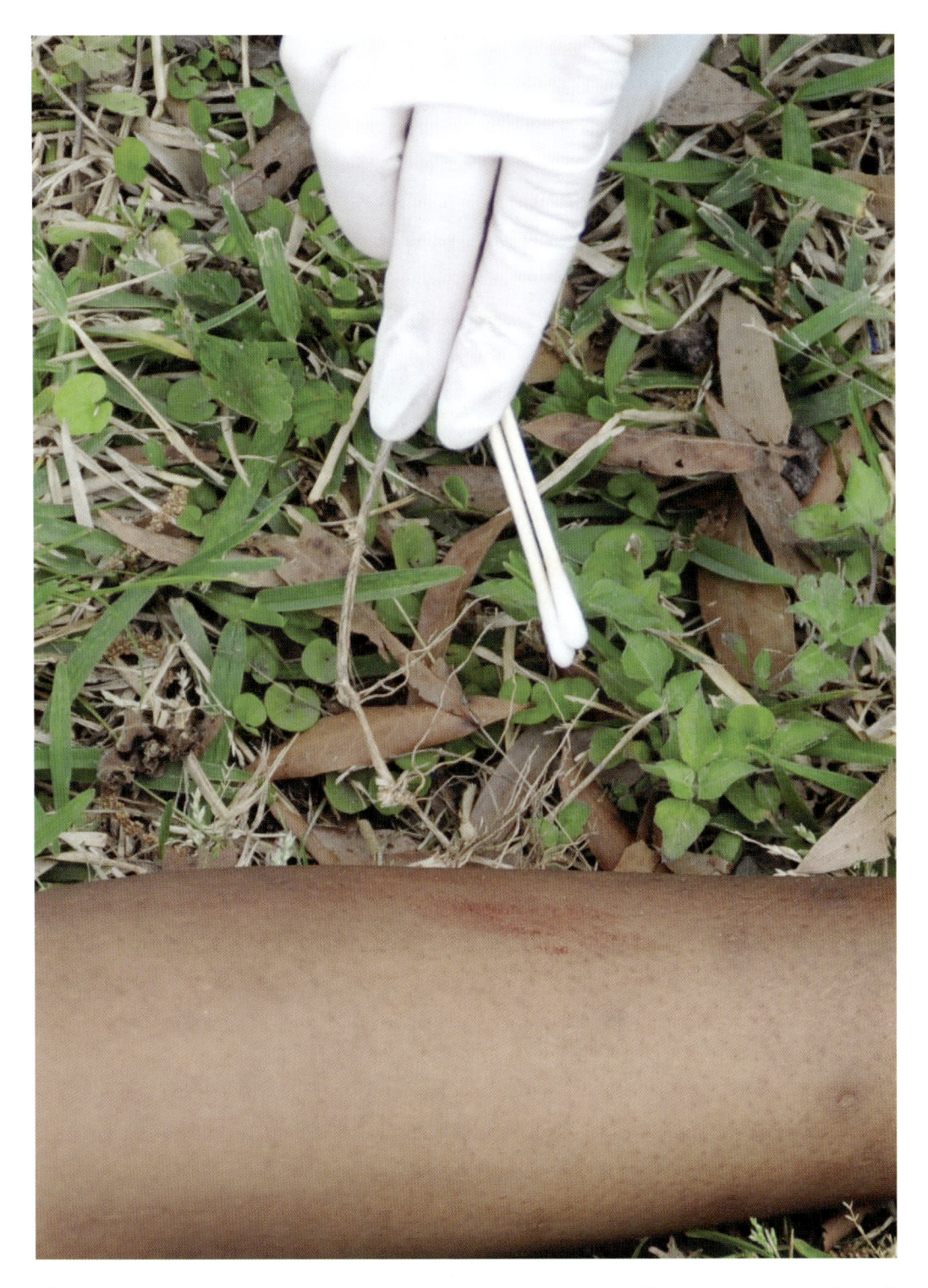

Figure 5.7 The blood smear in this case looked to be in an odd position. Very rarely will someone reach down to swipe his or her lower calf. That is why this stain stood out as unique and was swabbed. Any bloodstains that are found in unique locations or 90 degree blood drops should be swabbed.

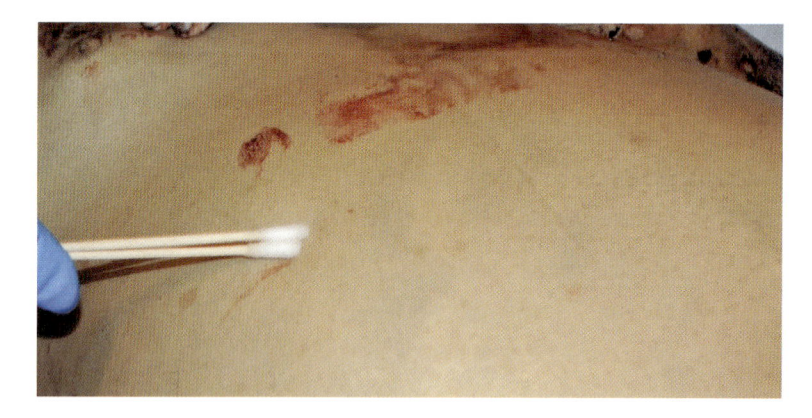

80 Forensic DNA Collection at Death Scenes: A Pictorial Guide

Figure 5.8 Blood smears on the body may have been made by the assailant. These smears should be swabbed and extracted for foreign DNA.

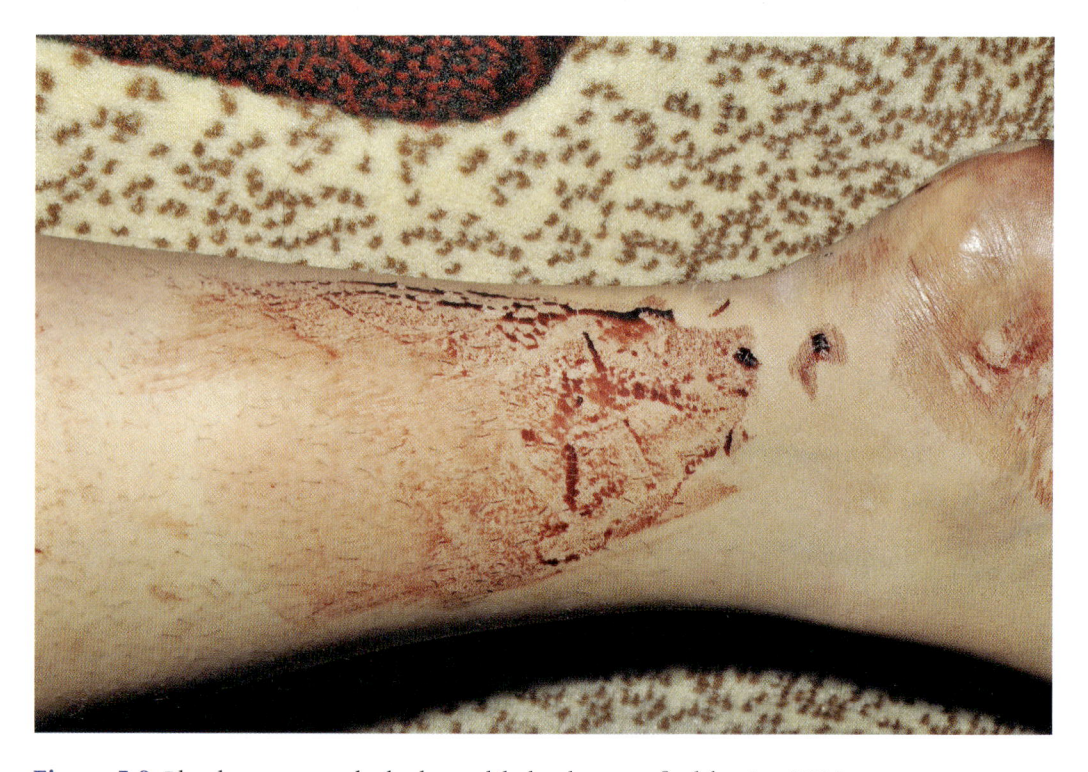

Figure 5.9 Blood smears on the body are likely places to find foreign DNA.

Blood Patterns 81

Figure 5.10 Blood prints found on clothing should be preserved for comparison with the object presumed to have made the print. We have found shoeprints, tire marks, and handprints in blood on clothing. The prints are cut from the clothing and preserved for further testing.

Figure 5.11 Blood patterns can guide the TECT analyst's collection. It appears the body was dragged across the carpet, and given the amount of blood on the upper portion of the body, the assailant may have held the body by the ankles to drag it. This led the trace analyst to swab the ankles to look for foreign DNA.

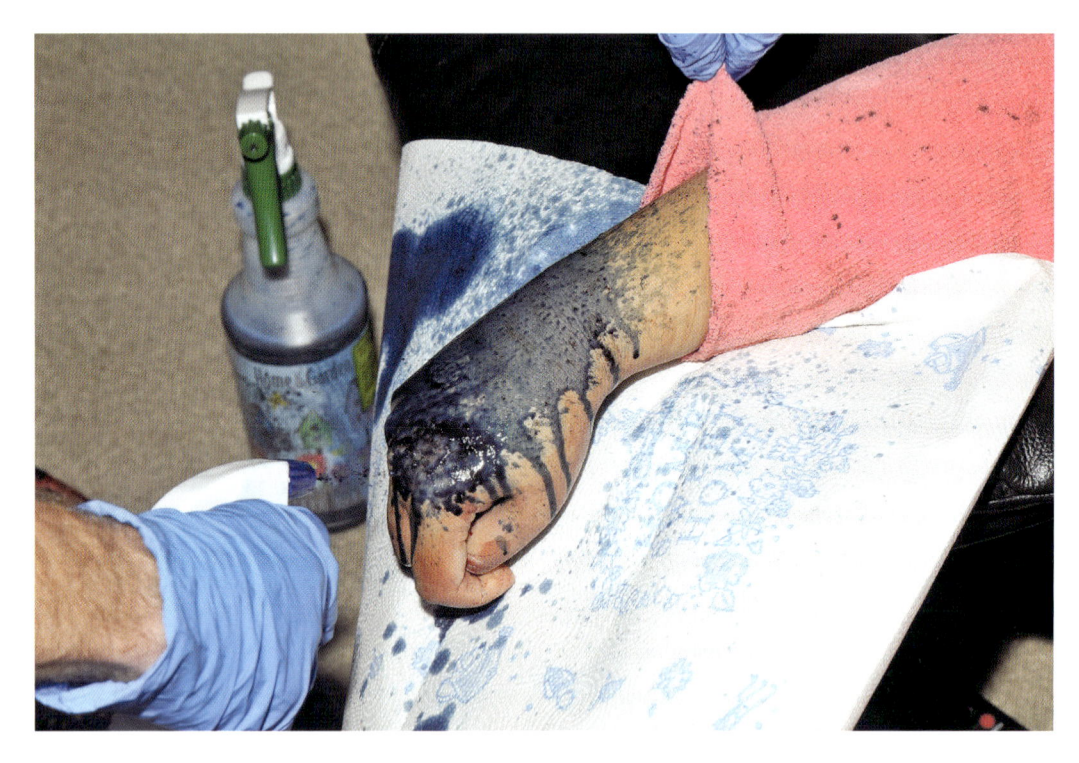

Figure 5.12 Amido black staining can be performed by CSU personnel on bloodstains when fingerprints appear to be present. After trace collection (sparing the area of the putative print), staining is performed. After staining and photography of the print, the print can be swabbed for DNA. Amido black will not interfere with DNA analysis.

Blood Patterns 83

Figure 5.13 This picture shows the results after amido black analysis. Amido black brings out the ridge detail of a fingerprint, even those on a body.

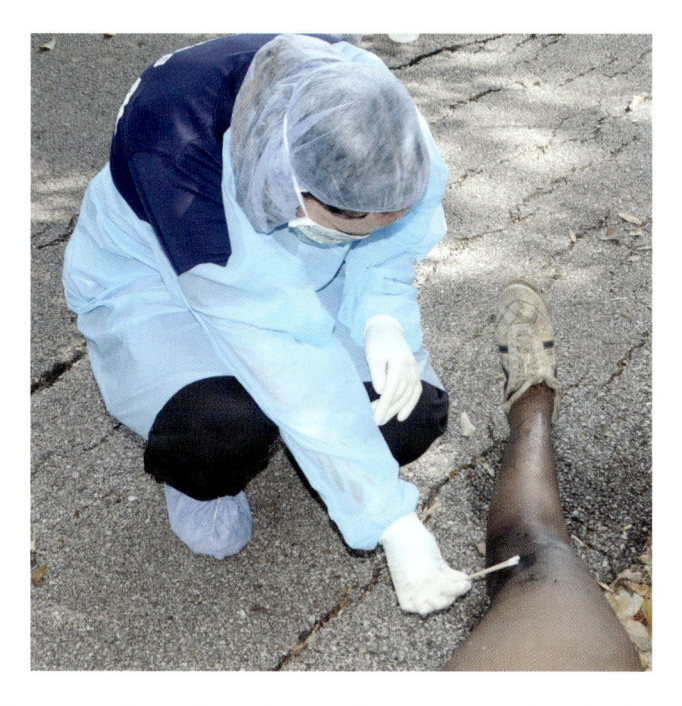

Figure 5.14 This picture shows the analyst swabbing stained prints developed by amido black stain. The print may produce touch DNA profiles on the body.

Bindings, Clothing, Wrappings

6

The TECT has had marked success recovering foreign DNA from bindings on decedents. Examples of common bindings are duct tape, zip ties, handcuffs, rope, and electrical cords. Foreign DNA has been recovered from each of these types of bindings.

The decedent's clothing can also be a good source of foreign DNA. An assailant may grab or rub the decedent's clothing, causing epithelial cells to be trapped there. Turned-out pants pockets have proven to be an excellent source of foreign DNA from decedent clothing. High levels of foreign DNA have been found in approximately 82% of every pocket swabbing tested—earning such turned-out pockets the moniker "hot pockets." Pockets produce profiles ranging from a few alleles below threshold to profiles that meet eligibility requirements for the Combined DNA Index System (CODIS) to complete DNA profiles (Moser, 2013).

Body wrappings are very helpful in preserving evidence for trace collection. Hairs, fibers, or DNA can be trapped and preserved within the wrapping material. Upon arrival at the scene, the TECT analyst will first search for and collect evidence from the outside of a wrapping (e.g., a comforter) before the wrapped body is placed in a sealed body bag. In the morgue the body will be unwrapped layer by layer while evidence is collected from each layer. The body will then be examined for any trace evidence. Finally, the wrappings are packaged and submitted to the investigating agency for further laboratory examination (Figures 6.1 to 6.42).

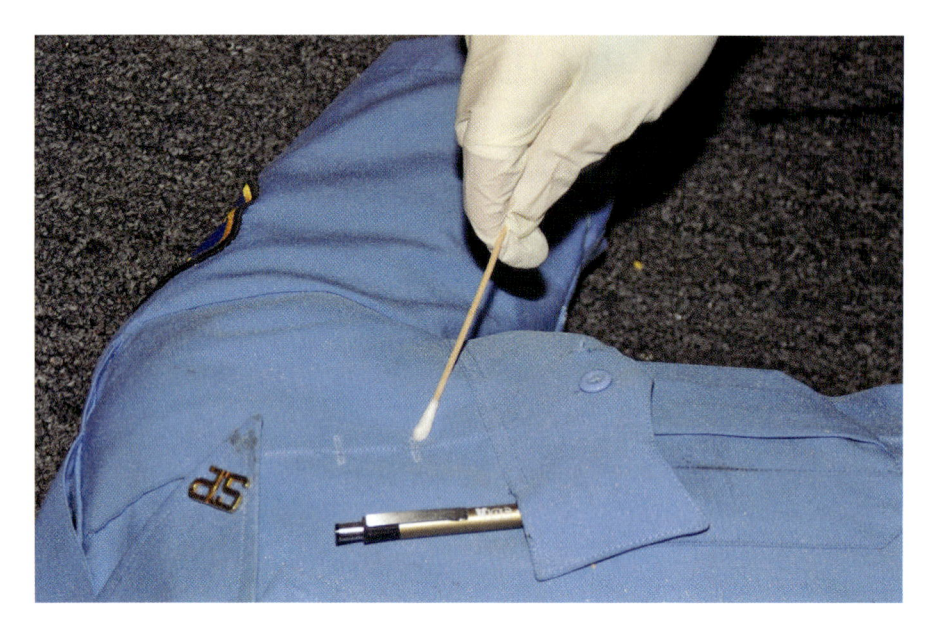

86   Forensic DNA Collection at Death Scenes: A Pictorial Guide

Figure 6.1 In this case the badge was ripped off of the officer. Swabbings were taken around the area of the missing badge for touch DNA. Looking for the disturbed objects at the scene and on the body is the best way to identify areas where foreign DNA may be present.

Figure 6.2 In this case the officer's weapon was apparently removed by the assailant; the decedent was shot in the head with his own weapon. Knowing this prior to the collection was helpful, as the analyst knew that touch DNA could be present around the weapon holster. This area was swabbed for foreign DNA.

Bindings, Clothing, Wrappings 87

Figure 6.3 Voided areas without blood, torn areas on clothing, and areas that look to have been pulled should be swabbed for foreign DNA.

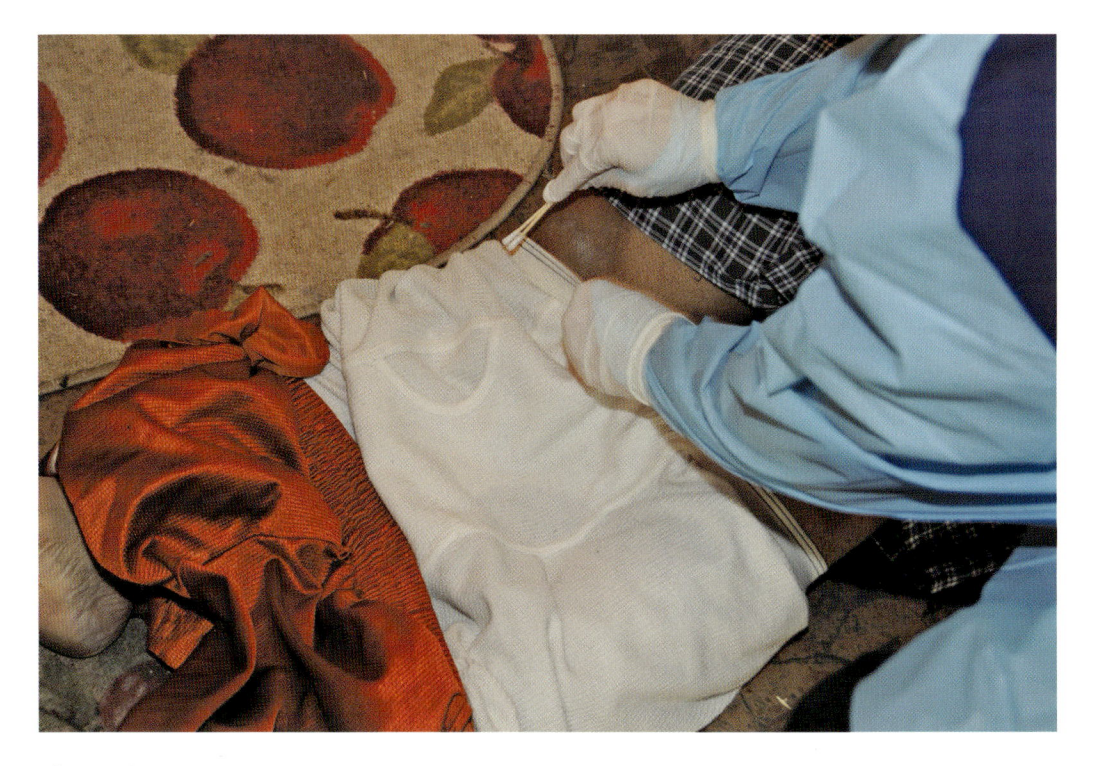

Figure 6.4 If the decedent is found with pants pulled down, the waistband should be swabbed for foreign DNA. Note that the decedent may have been sexually assaulted.

88　　Forensic DNA Collection at Death Scenes: A Pictorial Guide

Figure 6.5 If the decedent is wearing gloves, swab the outside of the glove. Gloves can capture DNA cells when there was contact between the assailant and decedent. Gloves should be swabbed just as the hand of the decedent would be.

Bindings, Clothing, Wrappings 89

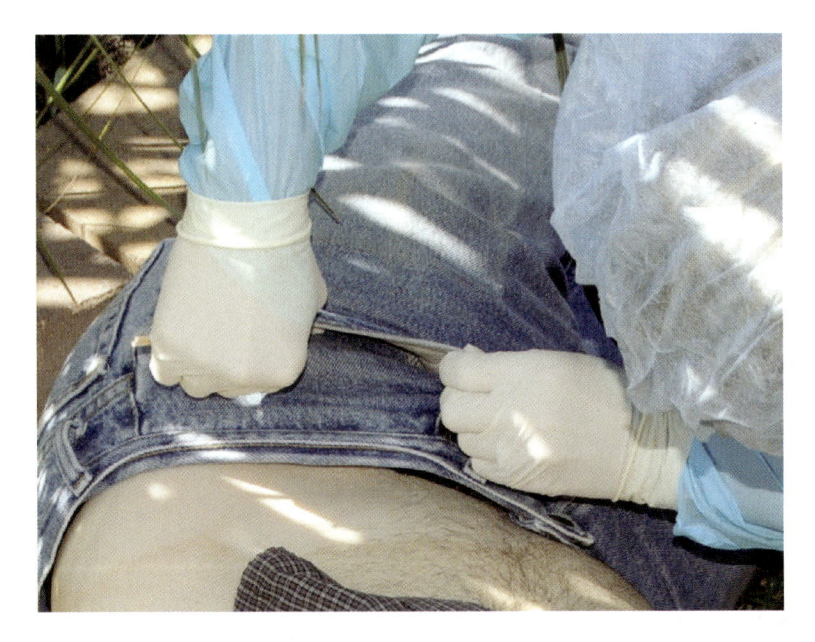

Figure 6.6 At times, the pockets are not turned out when objects are taken from them. If the decedent was known to be carrying something in a pocket that is missing, consider swabbing the pocket.

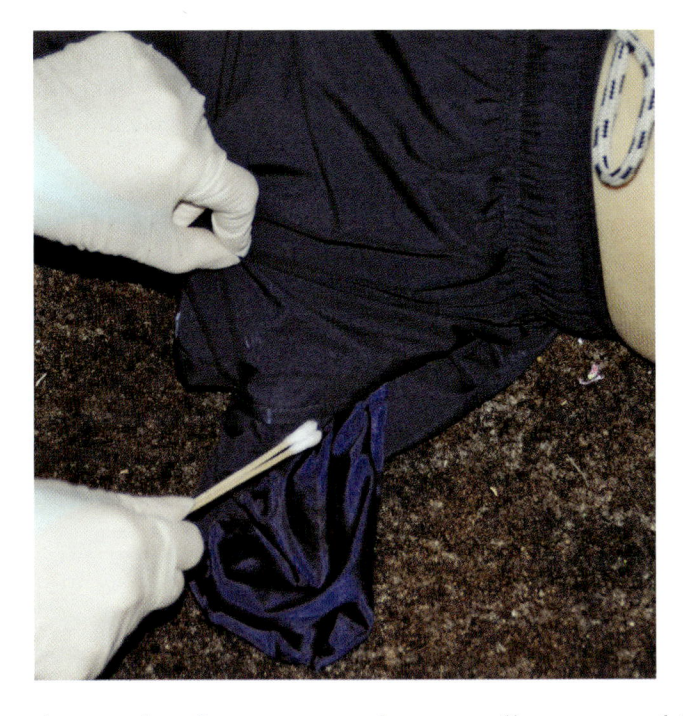

Figure 6.7 Turned-out pockets have proven to be an excellent source of foreign DNA. It seems intuitive that the inside of a pocket would yield DNA primarily from the owner of the pants. However, in our experience, inside-out pockets frequently yield more foreign alleles than decedent alleles. Turned-out pockets should be swabbed whenever they are observed at a crime scene.

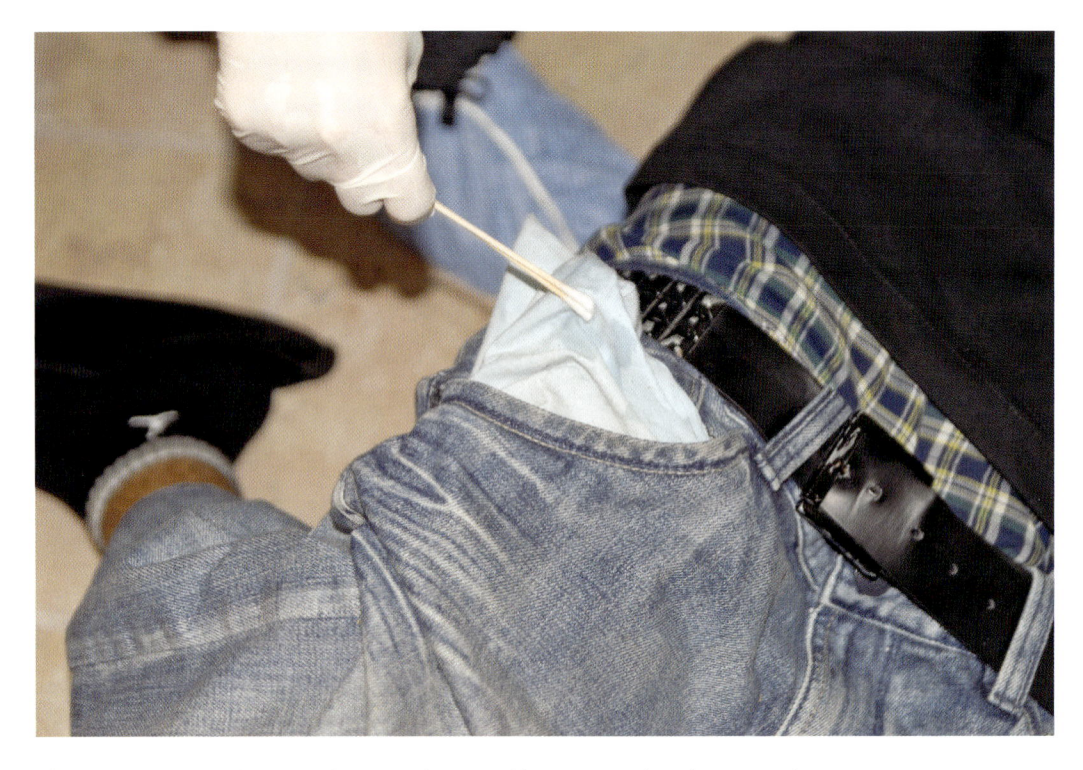

Figure 6.8 Turned-out pockets are key swabbing areas for obtaining foreign DNA.

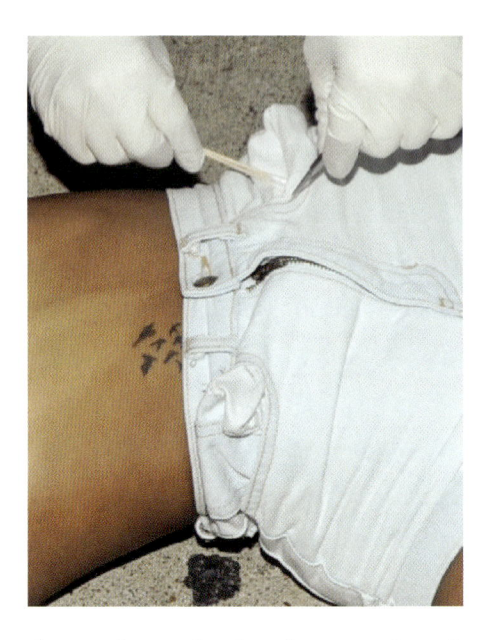

Figure 6.9 Turned-out pockets indicate the decedent may have been robbed. These pockets are swabbed looking for the last person who came in contact with the decedent and his or her pockets.

Figure 6.10 If blood is present on a turned-out pocket, it should be swabbed in void areas, separately from bloody areas. The pocket can be cut out and preserved prior to transport of the body for a more thorough analysis in the lab.

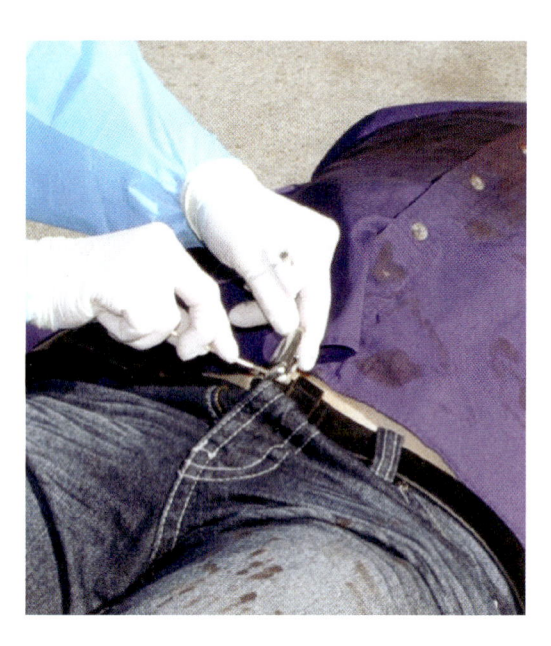

Figure 6.11 If a belt and button of the decedent's pants are undone, the assailant may have touched this area. It should be swabbed for foreign DNA.

Figure 6.12 Undone belts and buttons should be swabbed for foreign DNA.

Bindings, Clothing, Wrappings 93

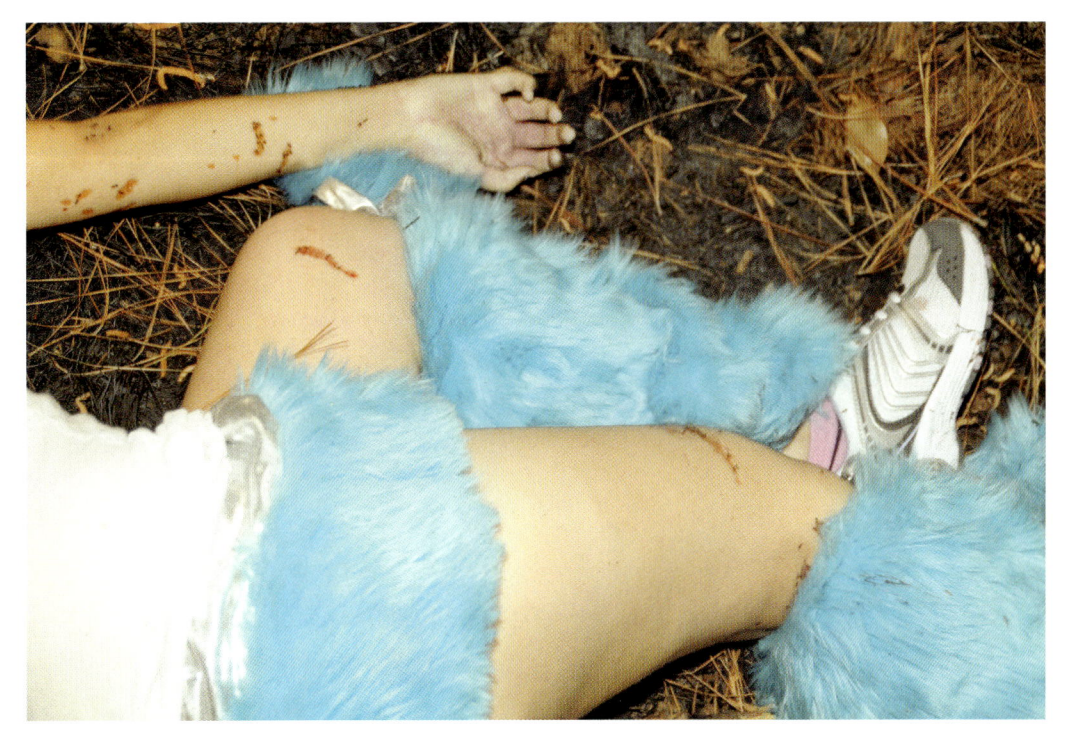

Figure 6.13 In some cases, unusual fibers may be helpful. If the assailant has transported the body, fibers from the decedent's clothing might be used to compare with fibers found in a suspect's vehicle, for example.

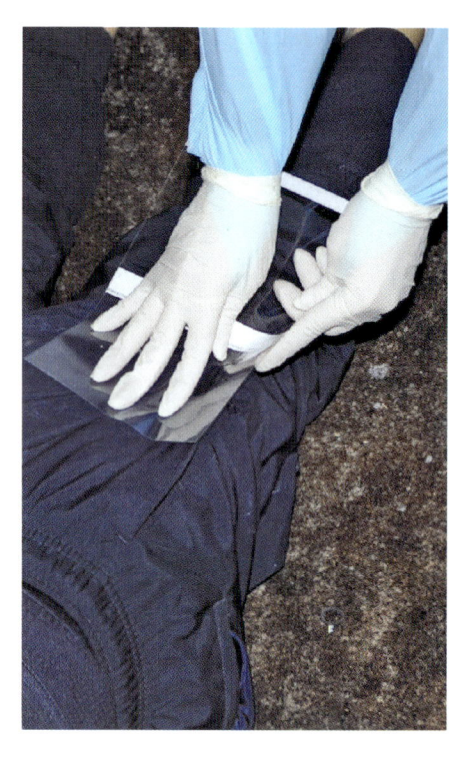

Figure 6.14 Clothing may contain trace evidence, such as hairs, fibers, or touch DNA. Tape lifts will pick up these items for processing in the laboratory.

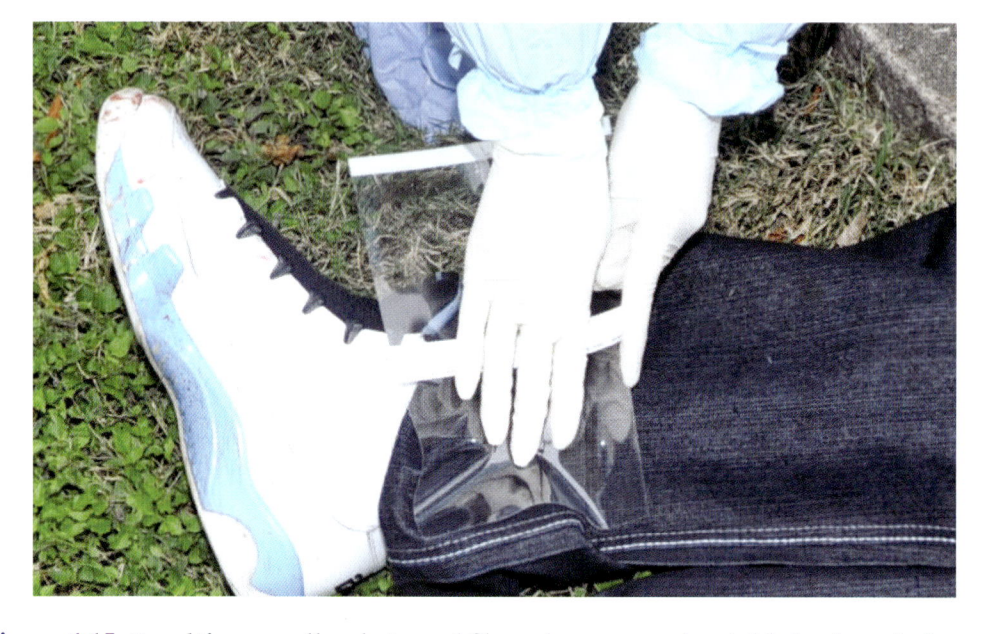

Figure 6.15 Tape lifts can collect hairs and fibers that may not be visible by the naked eye.

Bindings, Clothing, Wrappings

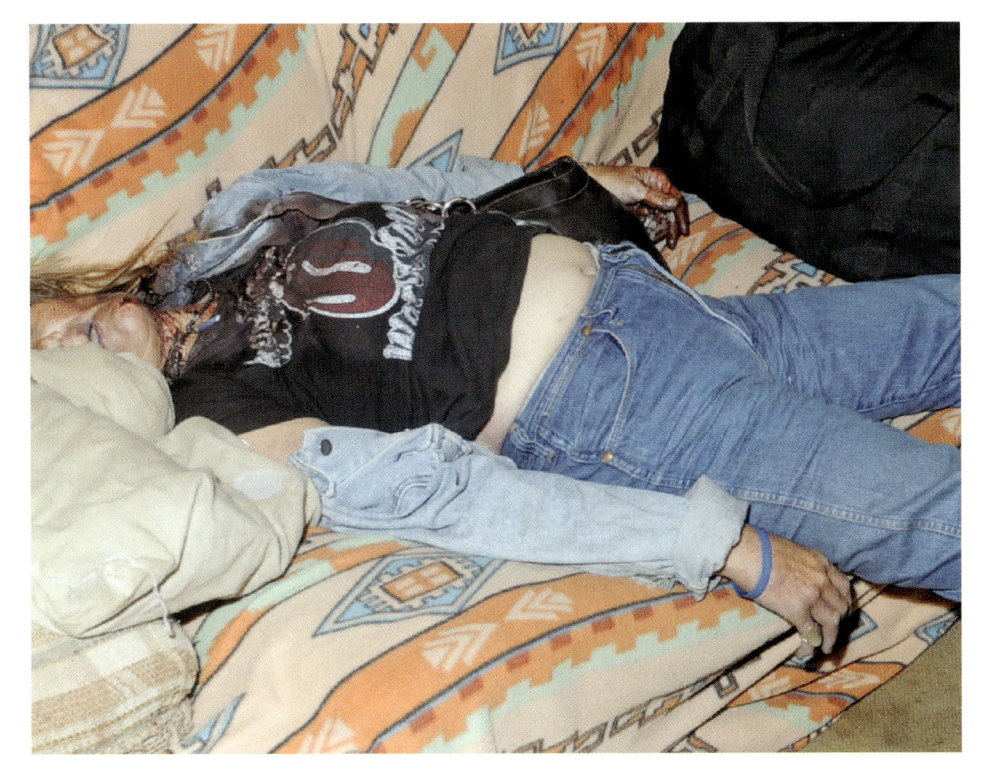

Figure 6.16 At some scenes, the decedent has no obvious areas to be swabbed. In this case, tape lifts of the entire body can be processed to look for foreign DNA. In the lab, the tape lift is swabbed with water to remove DNA that is on the lift.

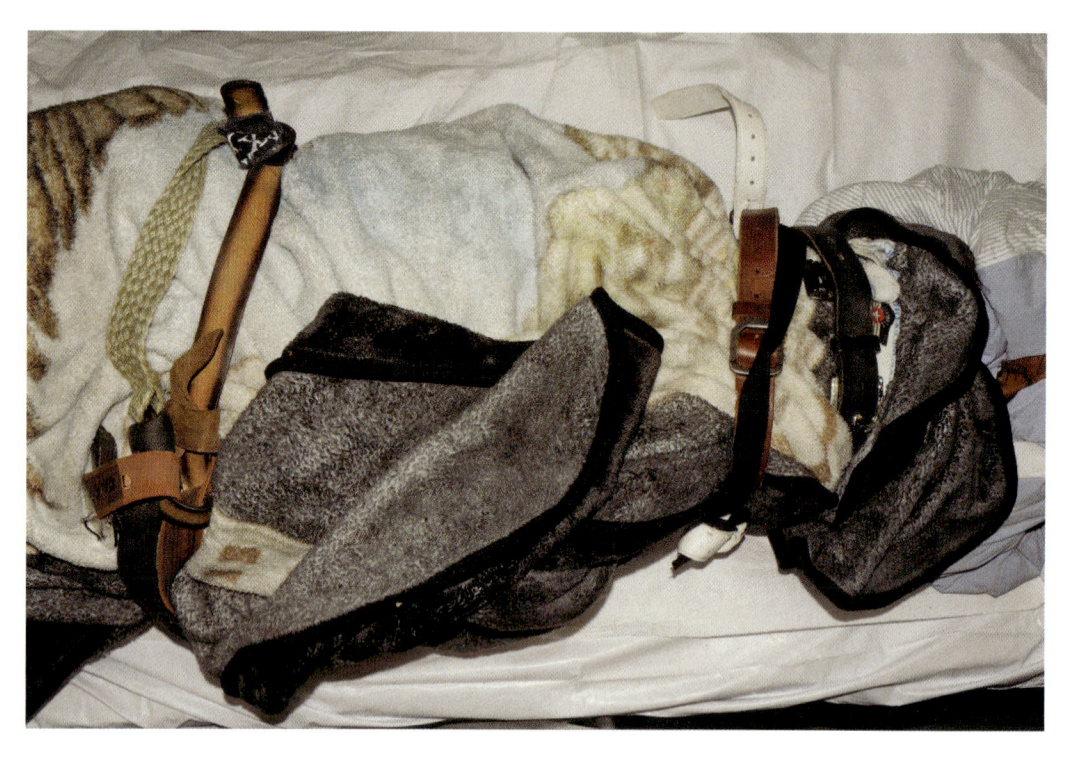

Figure 6.17 The decedent was bound with several belts and material. Each belt was fastened tightly. Each belt was swabbed in the buckle area and, separately, along the length of the belt. It was not known to whom the belts belonged, so both areas of each belt were swabbed.

Bindings, Clothing, Wrappings 97

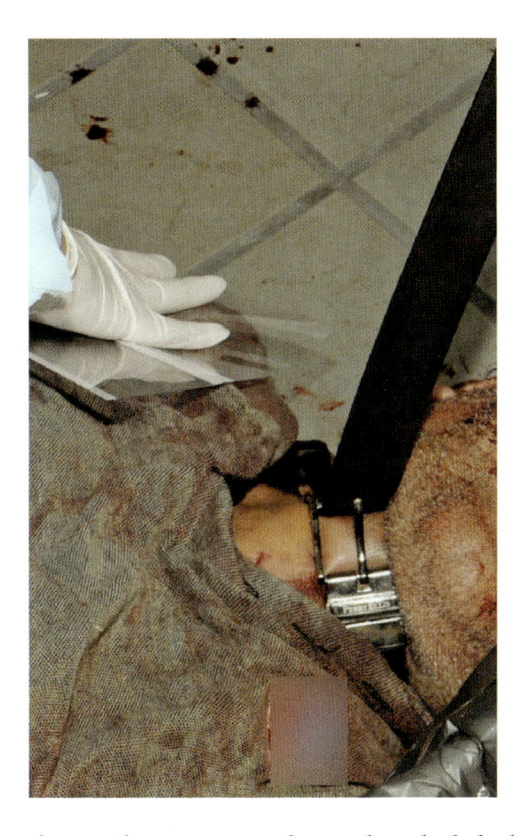

Figure 6.18 In this case, the assailant appears to have taken the belt the decedent was wearing and used it to strangle him. Accordingly, the assailant is likely to have touched it, perhaps with ungloved hands. The ends of the belt and the buckle area were swabbed to look for foreign DNA.

Figure 6.19 Duct tape is a useful source for trace evidence. It can hold hairs, fibers, and touch DNA, and it can capture fingerprints. Duct tape must be protected to preserve the trace evidence. After fingerprint analysis, the duct tape can be submitted to the laboratory for DNA analysis.

Bindings, Clothing, Wrappings 99

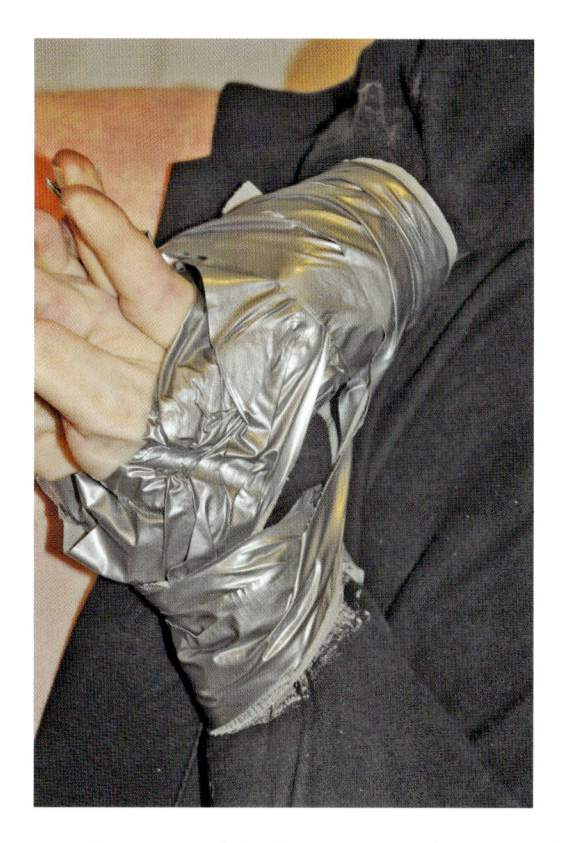

Figure 6.20 Duct tape must be preserved for fingerprint and DNA analysis.

Figure 6.21 Shoelaces are also encountered in bindings. The knot in a shoelace should be swabbed, and the area between the knots can be swabbed separately or swabbed using a gauze pad.

Bindings, Clothing, Wrappings 101

Figure 6.22 In some cases, a decedent is bound, apparently to excess, with duct tape and electrical cords. In this instance, the tape should be processed separately from the cords. The cords can be swabbed around the knot area and a gauze pad can be used on the long stretches of cord in between the knots. Duct tape should be cut off and preserved for fingerprint analysis and then submitted to the laboratory for DNA testing.

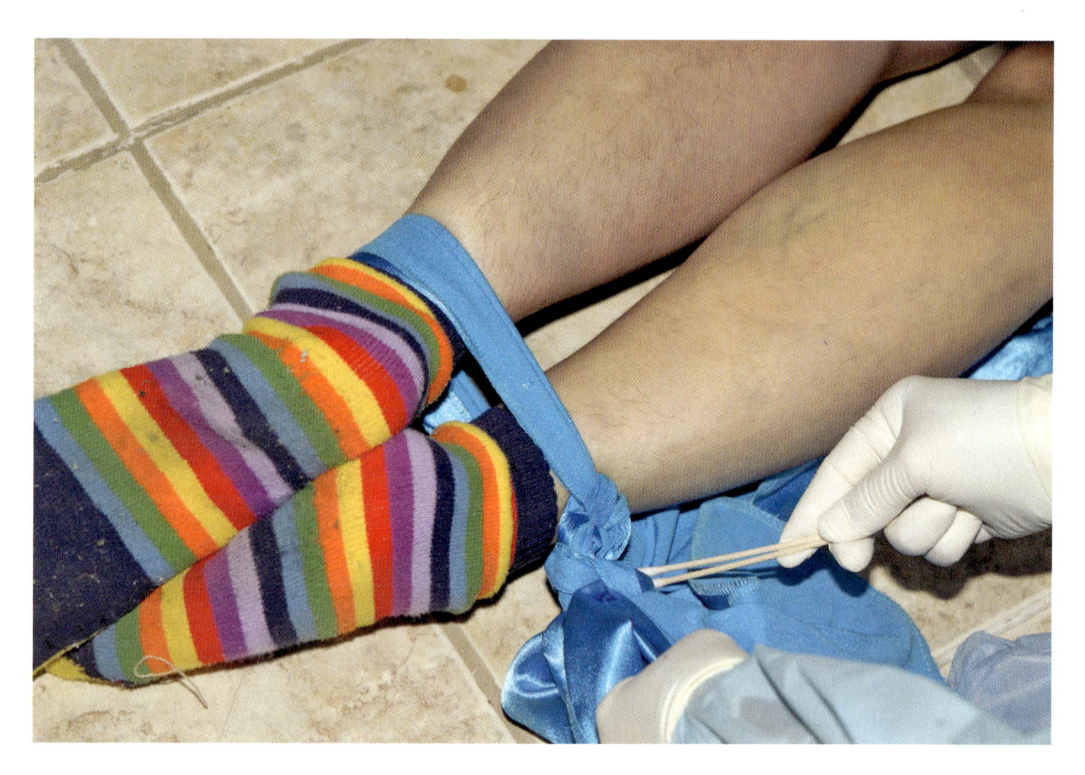

Figure 6.23 Cloth bindings are also encountered as restraints. The knots in cloth bindings should be swabbed for foreign DNA.

Bindings, Clothing, Wrappings 103

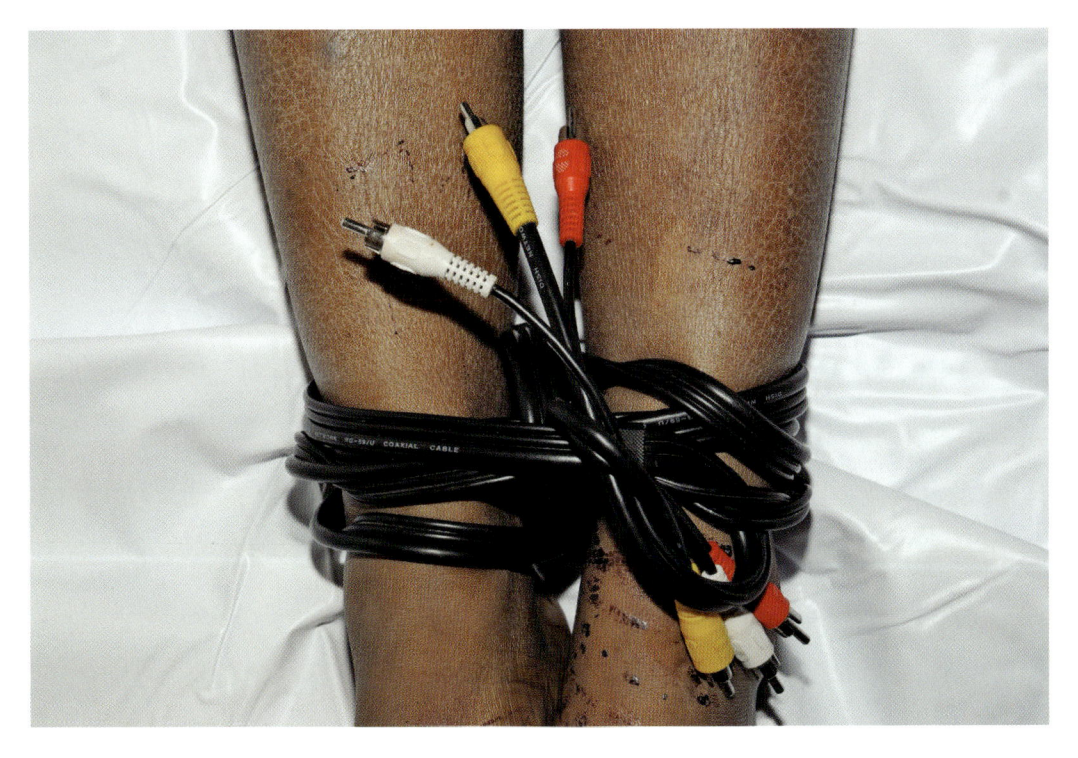

Figure 6.24 Cords are often used to bind individuals. They should be swabbed in the locations of knots or ties, and swabbed separately in the areas between the knots.

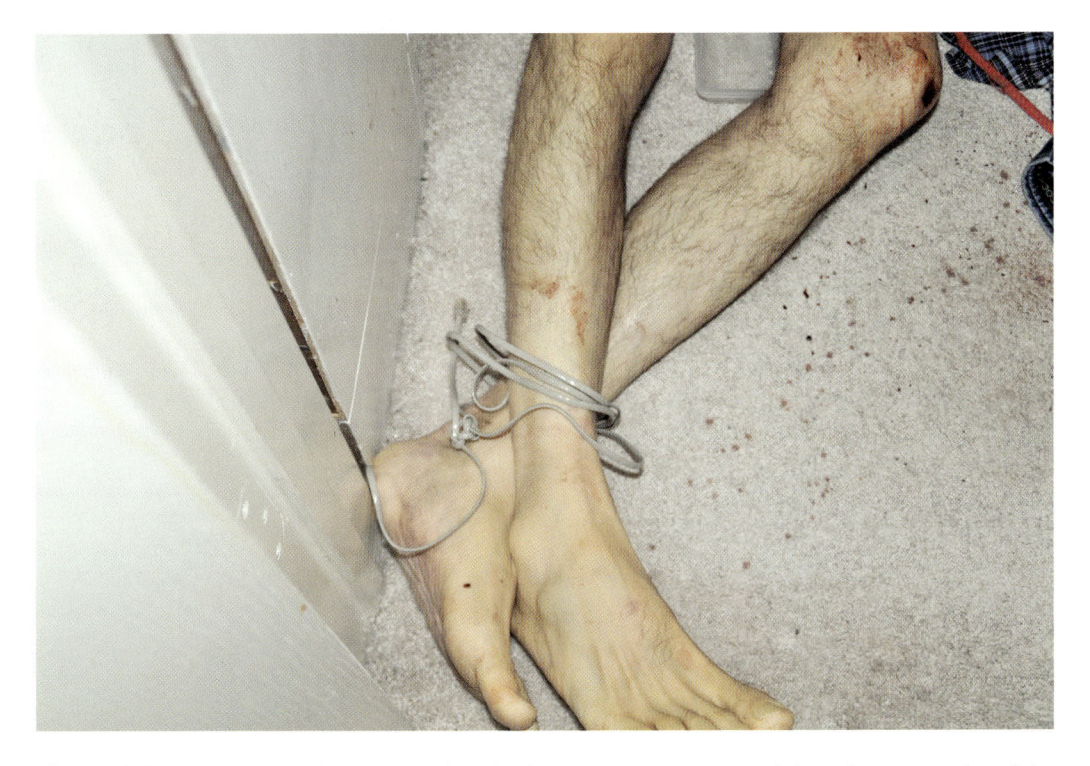

Figure 6.25 Cords should be swabbed at the knot or tie junction and then the remainder of the item swabbed separately.

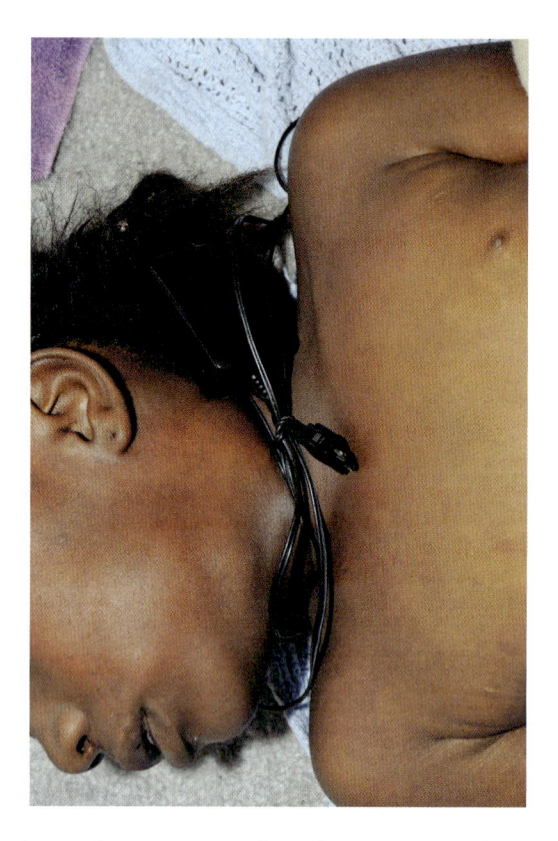

Figure 6.26 Some cords may have an area where fingerprint analysis can be performed. This area should not be swabbed and should be preserved for fingerprints and DNA analysis.

Bindings, Clothing, Wrappings 105

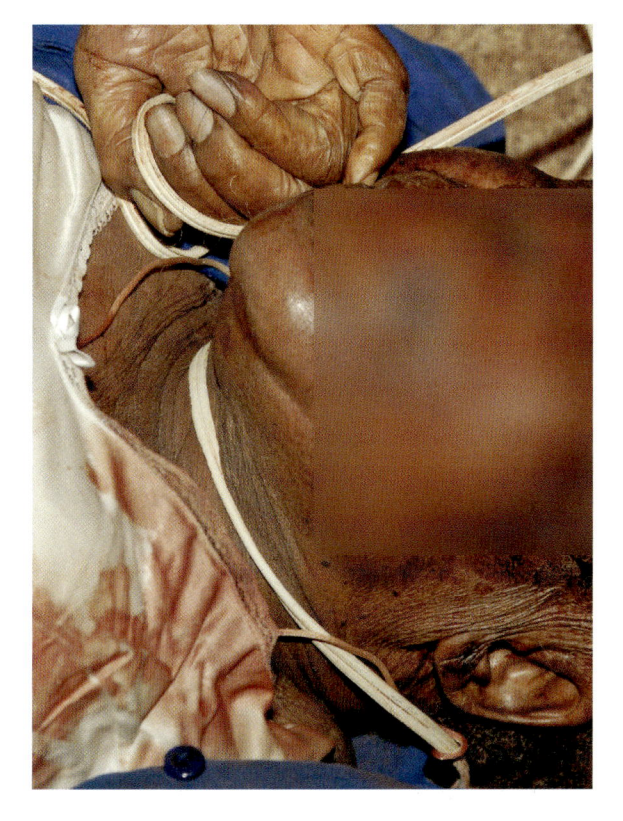

Figure 6.27 If corded areas are tied into knots, they can be swabbed to obtain DNA that may be present. If the cord is extremely long, a single swab may become saturated with dirt and other foreign particles that impede absorbance. A gauze pad is recommended if the area being swabbed is large or too long to use a swab.

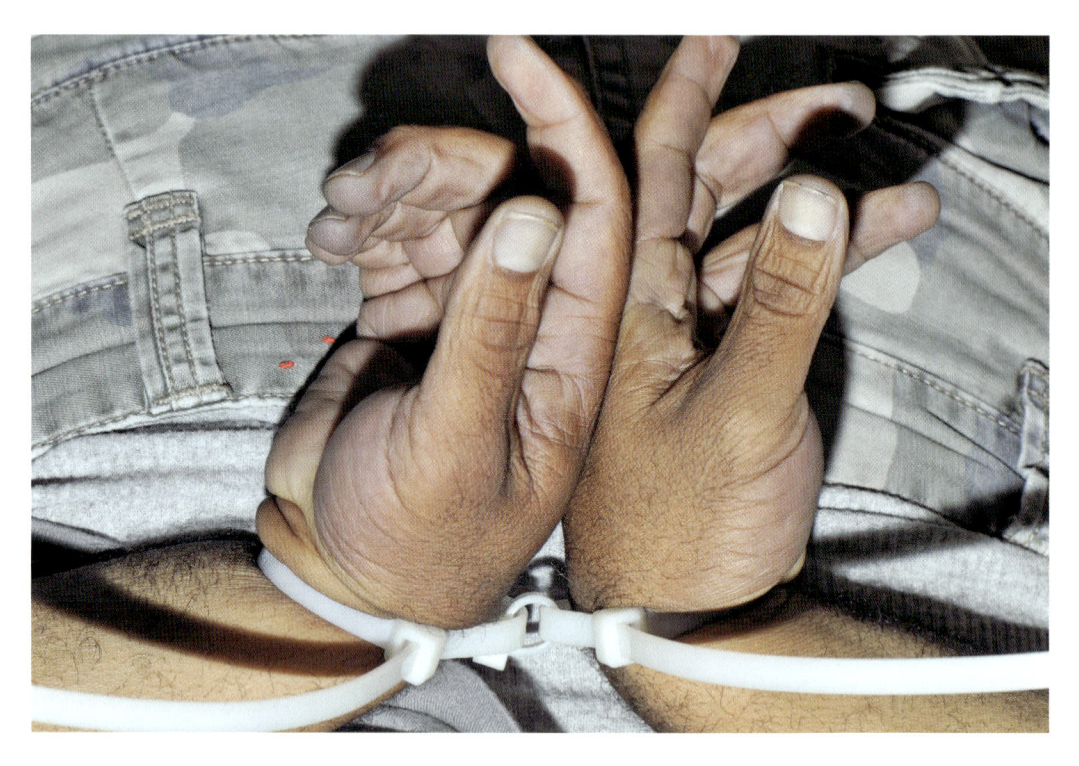

Figure 6.28 Zip ties indicate physical contact between the assailant and the decedent. Swab the zip tie while it is on the decedent, taking care not to swab the decedent. If possible, the zip ties should be cut from the body for submission to the DNA laboratory for more thorough swabbing of all sides.

Bindings, Clothing, Wrappings

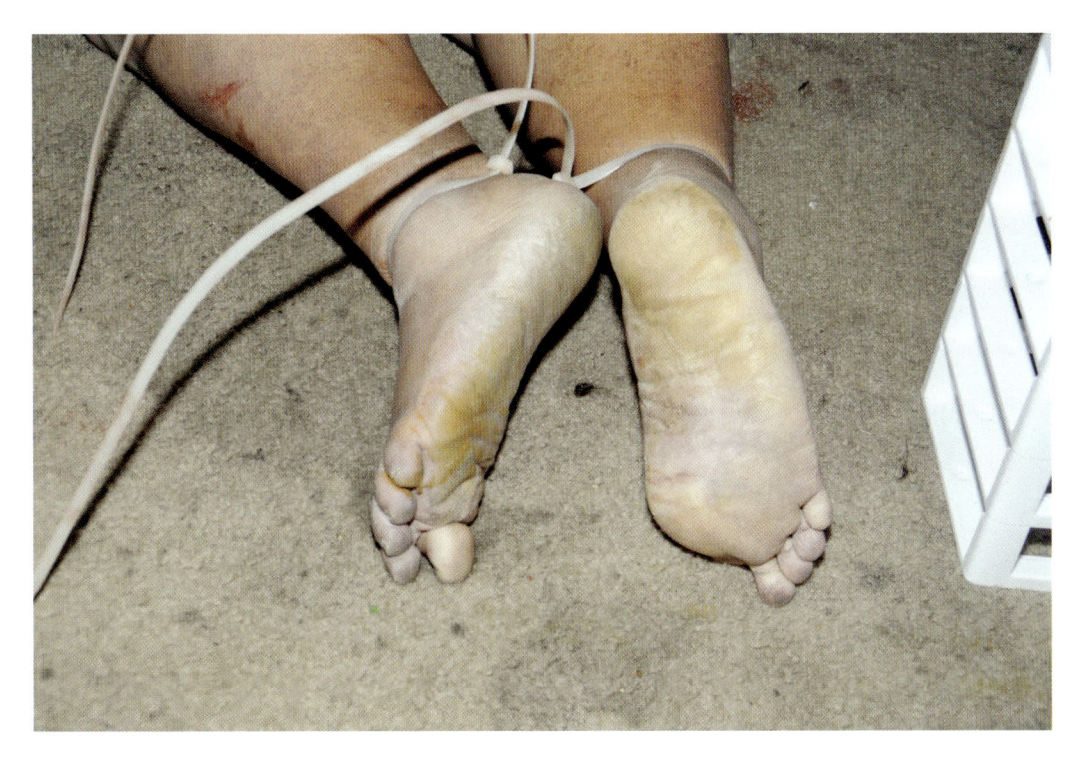

Figure 6.29 Zip ties should be swabbed because they indicate that physical contact has taken place.

Figure 6.30 Handcuffs as bindings are encountered at crime scenes. The shiny surface of the handcuff may be analyzed for fingerprints and should be printed prior to swabbing. The handcuffs can be protected in a paper bag and submitted for fingerprints and then DNA analysis.

Bindings, Clothing, Wrappings   109

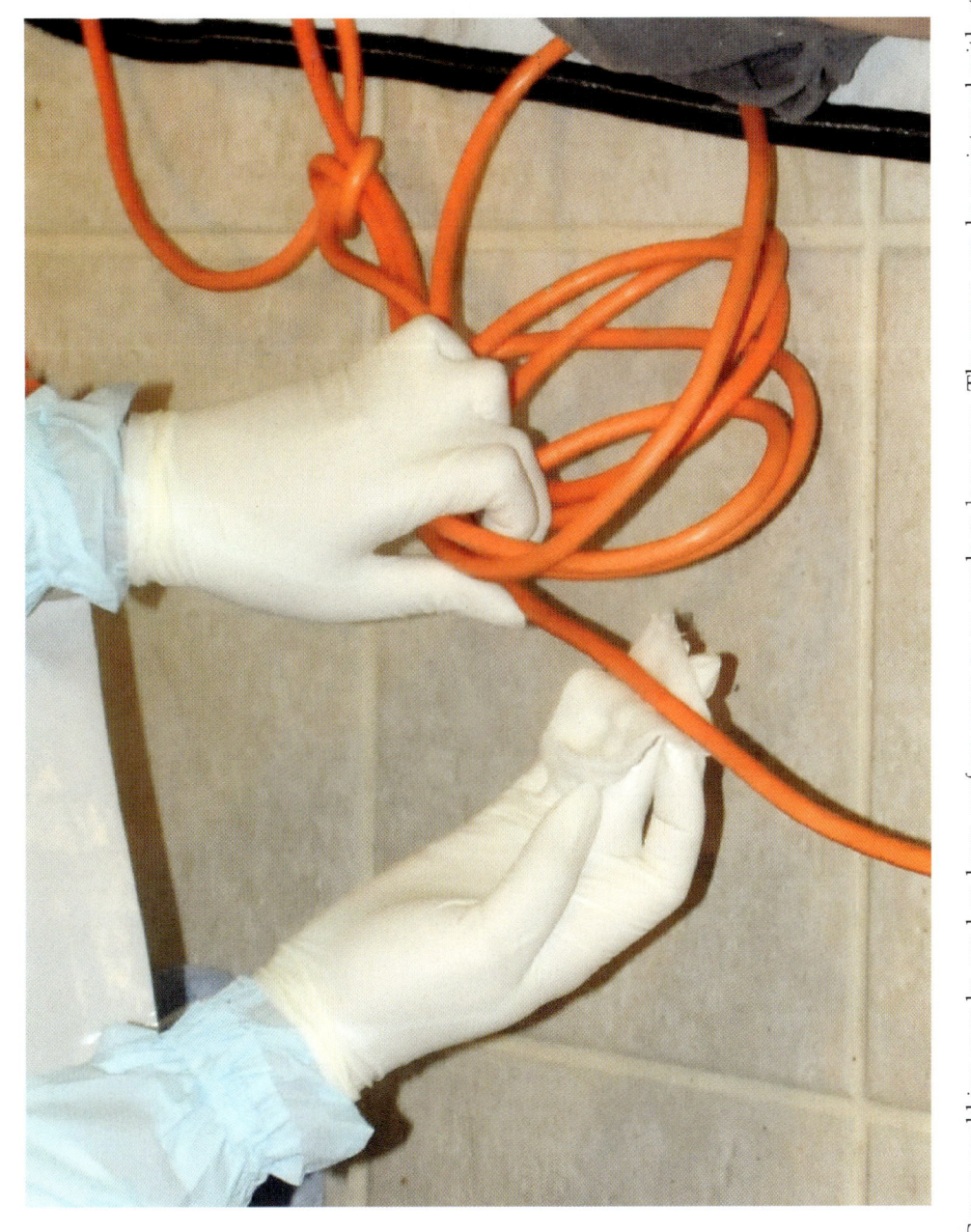

Figure 6.31 Gauze swabbing may be used on large surface areas or on cords or long rope. The gauze can be moistened with sterile water and wrapped around the cord and gently rubbed to collect DNA.

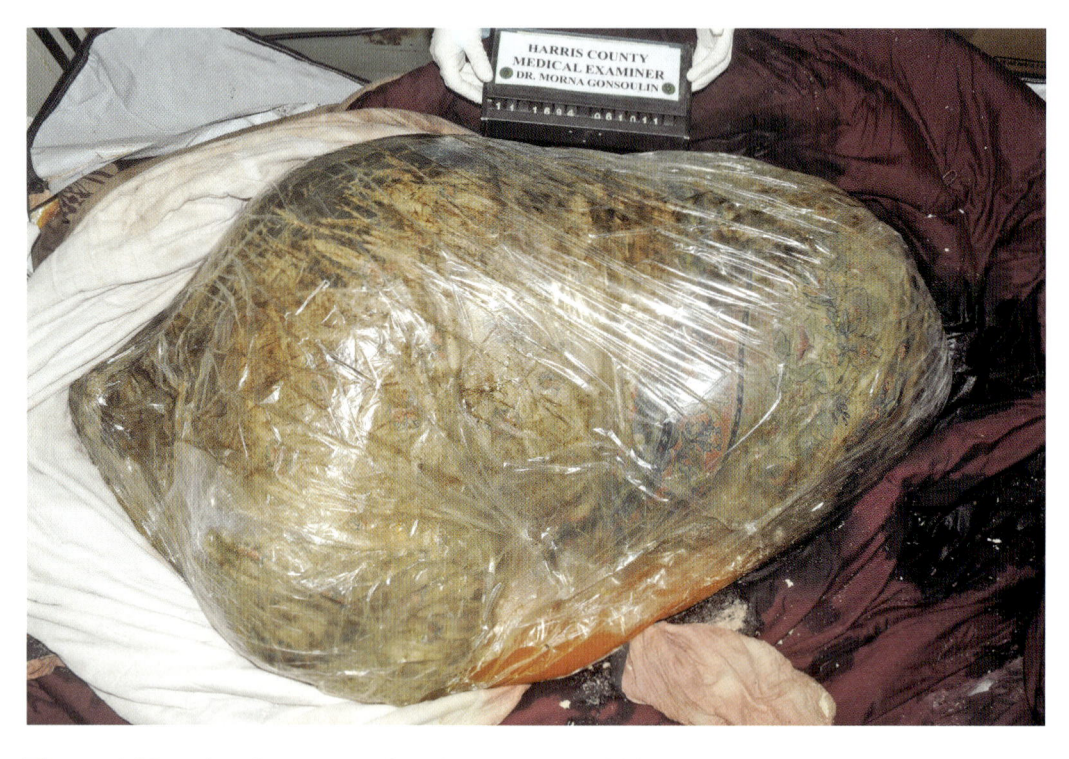

Figure 6.32 A decedent wrapped within items should be processed in a systematic manner. The outside of the wrapping should be tape lifted and swabbed at the scene. The next layer should be analyzed in the morgue under controlled conditions. The outer layer and the inner layers should be submitted to the DNA laboratory for DNA analysis.

Bindings, Clothing, Wrappings 111

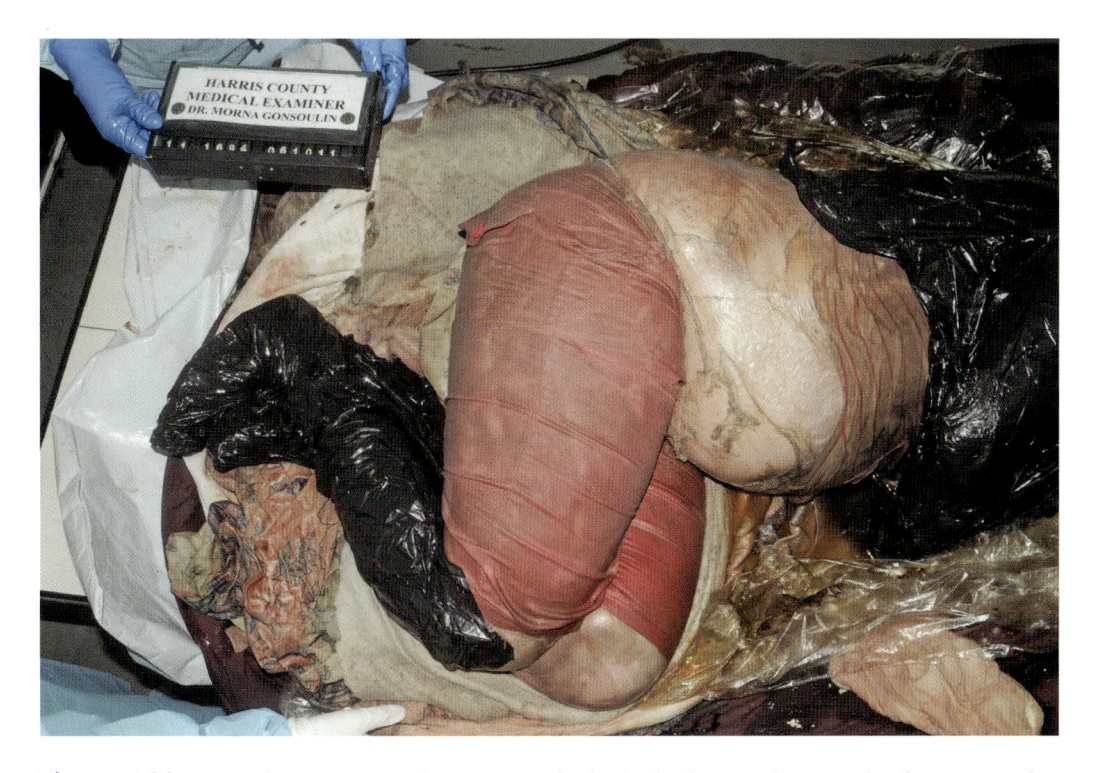

Figure 6.33 Once the innermost layer is reached, the body may be completely wet. In this instance, pickings should be performed on any hairs or fibers observed and then placed on a tape lift for preservation.

112 Forensic DNA Collection at Death Scenes: A Pictorial Guide

Figure 6.34 The photos show inner packaging that was opened and processed in the morgue. Each layer is swabbed and tape lifted to collect trace evidence.

Figure 6.35 The photos show inner packaging that was opened and processed in the morgue. Each layer is swabbed and tape lifted to collect trace evidence.

Bindings, Clothing, Wrappings 113

Figure 6.36 Knots and areas in a comforter that may have been touched on the outside should be swabbed. The comforter should then be transferred to the morgue and processed under more controlled conditions.

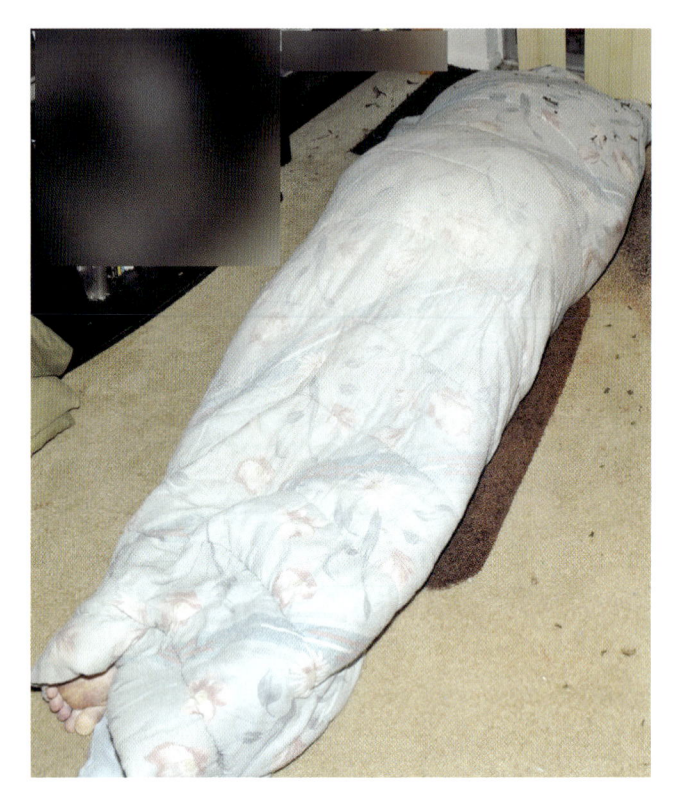

Figure 6.37 The outside of body wrappings can be tape lifted and knots can be swabbed. The body should not be unwrapped until present in the morgue under controlled conditions.

Bindings, Clothing, Wrappings 115

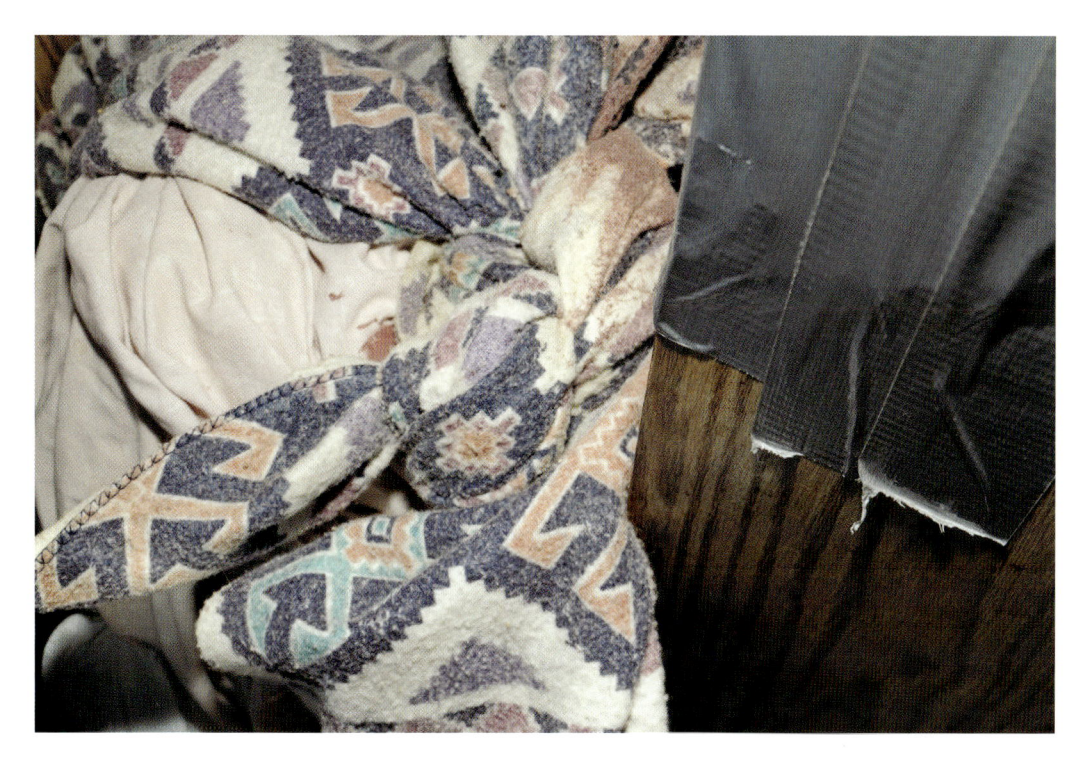

Figure 6.38 If knots are observed on a body wrapping, the suspect may have tied the knots with ungloved hands. The knots should be swabbed for foreign DNA.

Figure 6.39 If a decedent is found in a suitcase, the outside of the suitcase should be tape lifted and the handle and zipper should be swabbed. The entire suitcase should be transported to the morgue for additional collection in a controlled environment.

Figure 6.40 The outside of a suitcase should be swabbed around the handle areas and then tape lifted. The suitcase should not be opened until transported to the morgue, a more controlled environment.

Bindings, Clothing, Wrappings 117

Figure 6.41 Multiple steps are required to place someone in a rolled-up rug. Each step is an opportunity to leave DNA on the rug. The outsides and ends of the rug should be swabbed, and the rest of the rug should be tape lifted on the outside. The rug should be transported to the morgue for additional examination and collection in a controlled environment.

Figure 6.42 An alternate light source can be used to detect body fluids such as saliva or semen. This blue wavelength light source can help determine what areas of the body should be swabbed for stains not visible with the naked eye. Orange safety goggles must be worn to protect your eyes and for the stain to be visible.

Tough Places

7

Crime scenes can be challenging environments for TECT analysts. Some scenes have loaded weapons that are cocked or jammed. They can be very dangerous if intentionally or accidentally manipulated (Figure 7.1). Scenes may require analysts to work in very tight areas that are hard to access and difficult to maneuver in (Figure 7.2). Some scenes require wading into water, which could cause an analyst to lose his or her footing. Rubber boots are recommended to ensure that there is traction and to keep the analyst's feet dry. Wet scenes also present an electrocution risk, and the analyst must be cognizant of that hazard. Some scenes will require work in the middle of a hot summer day. It is important to bring water and to rehydrate while working a scene, especially while wearing a Tyvek® suit. Every scene is unique and challenging in its own right. To reduce loss of trace evidence, it is important to complete all possible collection activities before altering the position of or moving the body. However, safety must be the first priority at every scene (Figures 7.3 to 7.7).

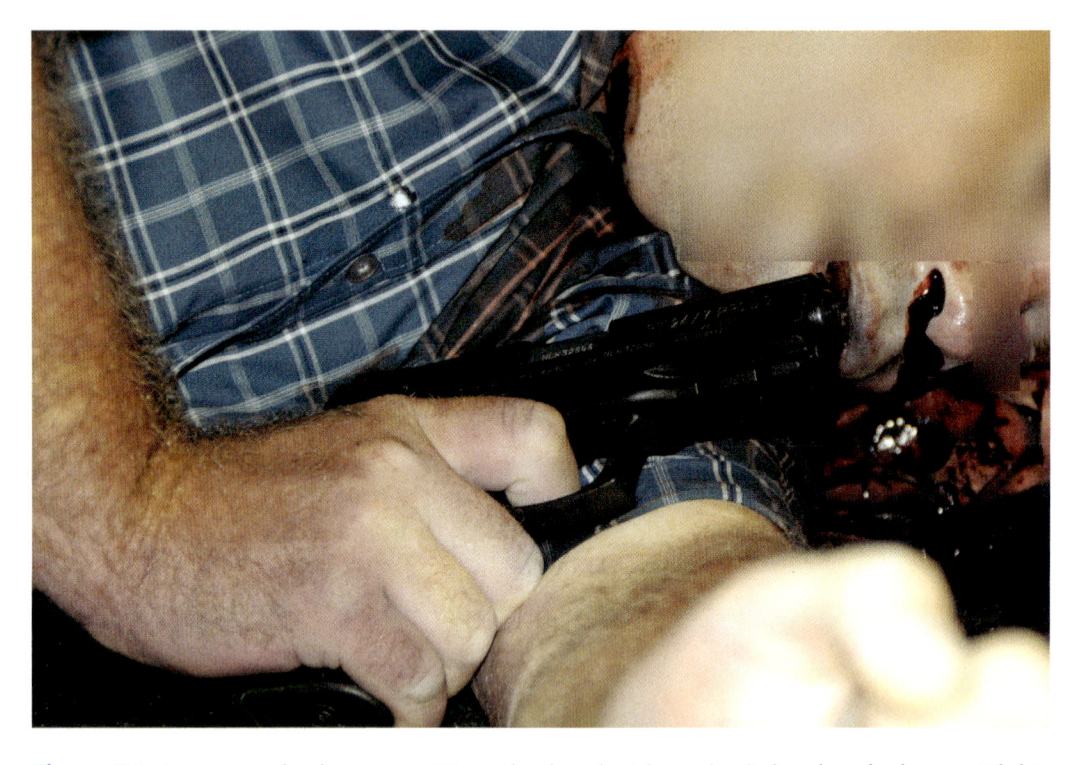

Figure 7.1 Scenes can be dangerous. Here, the decedent has a loaded and cocked gun with his finger on the trigger in rigor mortis in a car trunk. Movement of the body could cause the gun to fire. Law enforcement must make the gun safe and the scene ready for processing prior to any collection. At most trace scenes, no one may touch the body until the trace analysts have completed their collection. In this case, law enforcement was allowed to remove the gun while wearing gloves before the collection began.

Figure 7.2 Scenes can be in very tight places. The body is accessible from a limited number of angles in this trunk. The TECT analyst will process as much of the body as possible in that position in the trunk. The body will then be carefully placed into the body bag on the opposite side. Trace collection can then be performed on the other side while the body is secured in the body bag.

Figure 7.3 Scenes can present dangers such as broken glass, syringes, biological fluids, and drugs, among others. The analyst must be cautious when approaching a scene, touching nothing without protective gear. Press clothing tape lifts softly at first to ensure that nothing underneath is dangerous.

Tough Places

Figure 7.4 Bodies may be in tight spaces, here on the floorboard of a car. Maneuvering can be difficult, and it will be even more difficult if decomposition has begun. In this situation, approach through the opposite side of the vehicle, processing what you can reach. The body can then be moved into a body bag facing the opposite way. Processing can continue while the body is in the bag.

Figure 7.5 A body in a ditch presents additional challenges. In addition to limiting the analyst's ability to collect samples without disturbing the body, water may be present, making collection more difficult. It is best to process the body in the ditch without moving it. Rubber boots and a Tyvek suit are recommended for wading into the ditch. Once collection is complete on the accessible side, the body can be placed into a body bag on its opposite side, out of the ditch. A white light can be used to identify trace fibers or hairs that remain in the clothing. Of course, if the ditch looks too dangerous to enter, the body must be moved to a safe area before the collection can begin.

Tough Places   125

Figure 7.6 A body under a bridge can be problematic. Water may be present, complicating the collection. In this case, the concrete ditch wall is steeply angled, presenting a slip and fall hazard. A Tyvek suit is recommended along with rubber boots, if water is present. The body can be processed initially on the side facing up, and then placed inverted in a body bag in a flat, dry area to complete the collection. If the ditch looks too dangerous, the analyst must ask that the body to be moved to a safe location beforehand.

Figure 7.7 Here, a body found under a trailer presents a very limited work area. The analyst must first be confident the area under the trailer is safe. The analyst may begin by reaching under the trailer. Once all areas that can be reached have been processed, the body must be moved from under the trailer and placed in a body bag. Collection can then be completed from the bag.

Trace Success Cases

8

The trace team succeeds when evidence is recovered that may assist an investigation. Some successes link an identified suspect or an individual identified by a Combined DNA Index System (CODIS) match to the decedent. In other instances, the source of the foreign DNA is not immediately identified. Either way, the TECT analyst's work serves to protect evidence that might otherwise have been unknown or lost. The goal is to provide physical evidence to assist in the investigation of a death. As techniques of DNA analysis improve, the ability of the TECT analyst to provide DNA to assist an investigation will also improve.

Over the years, the focus of trace collections has shifted from fibers and hairs to touch DNA. In 2011, the success rate of tested trace evidence was 60%. These successes provided informative DNA results that led the investigator to the suspect or gave foreign DNA results that may be compared to a suspect at a later date. In 2012, 50% of the trace cases tested yielded foreign DNA results, useful in comparing to suspects in the case. DNA profiles range from a few alleles below threshold, suitable only for exclusion of a suspect, to profiles that meet CODIS eligibility requirements and result in a CODIS match. In mid-2013, we have already performed ~100 trace collections, double the number of collections performed for the entire year of 2012. There are many trace success cases being developed as we write this, and our hope is that the trend continues (Figures 8.1 to 8.25).

Figure 8.1 A female was abducted and found deceased in a field, her head wrapped with duct tape and a gunshot wound of the head through the tape. Because the bullet passed through the tape, it was not removed at the scene but swabbed while holding the decedent's head to prevent blood from contaminating it. Once collection was completed, the analyst held the decedent's head in the same position, also to prevent blood from contaminating the tape, while the body was placed carefully into a body bag. After autopsy, the investigating agency processed the tape and was able to identify a probative fingerprint on the sticky side that might have been damaged or lost without the TECT analyst's efforts to protect it.

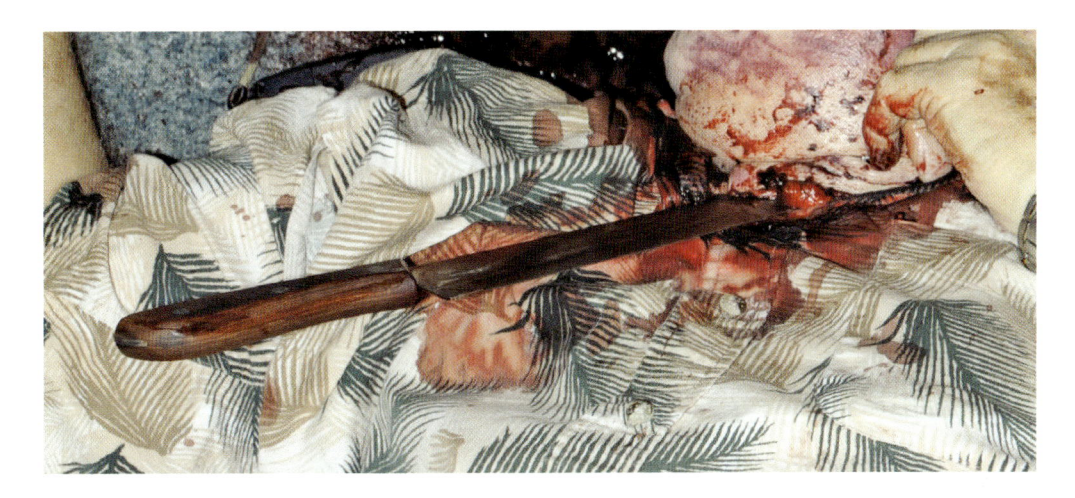

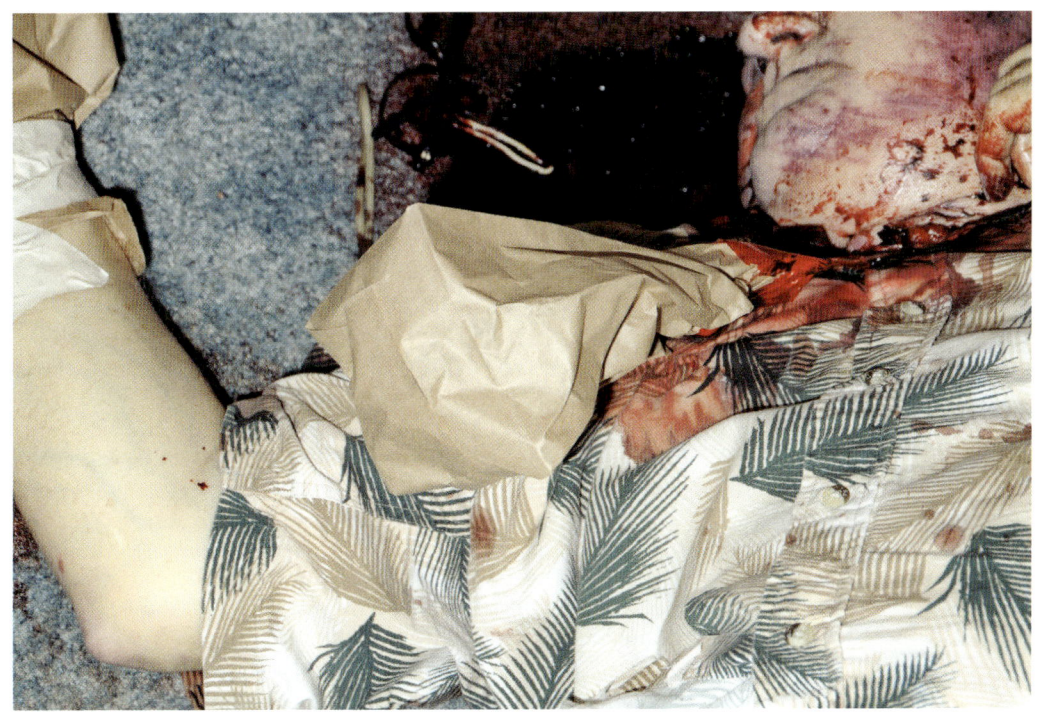

Figure 8.2 A male was found on his houseboat deceased with a knife in his neck. Since the knife had to remain in the body for the pathologist to examine at autopsy, the trace analyst protected the handle to prevent damage or loss of fingerprints and DNA during transport. The knife handle subsequently revealed a DNA mixture with a potential association to a suspect.

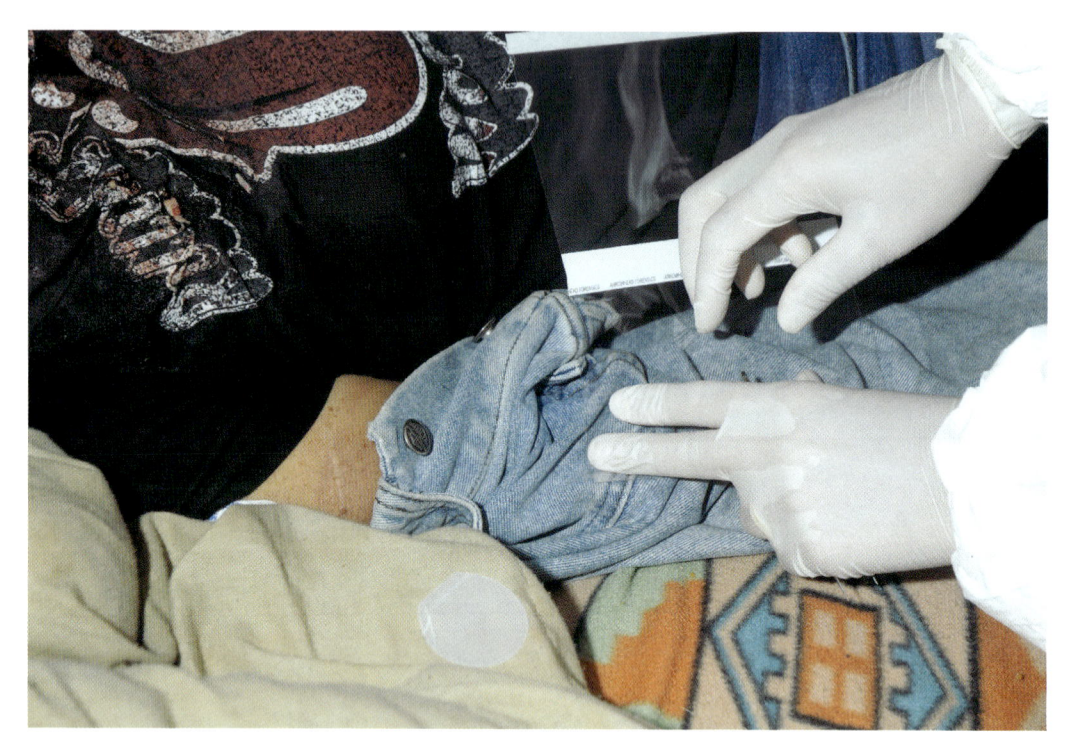

Figure 8.3 A female was found on the couch in her home deceased with multiple stab wounds. Tape lifts of the right sleeve of her clothing revealed a DNA mixture. The mixture linked the decedent and one of the suspects as possible contributors to the mixture. These results illustrate what we have seen repeatedly: informative DNA results can be obtained from tape lifts or swabs of clothing that was "grabbed," clothing that has no apparent staining. In some instances significant amounts of DNA are transferred when clothing is handled.

Trace Success Cases 131

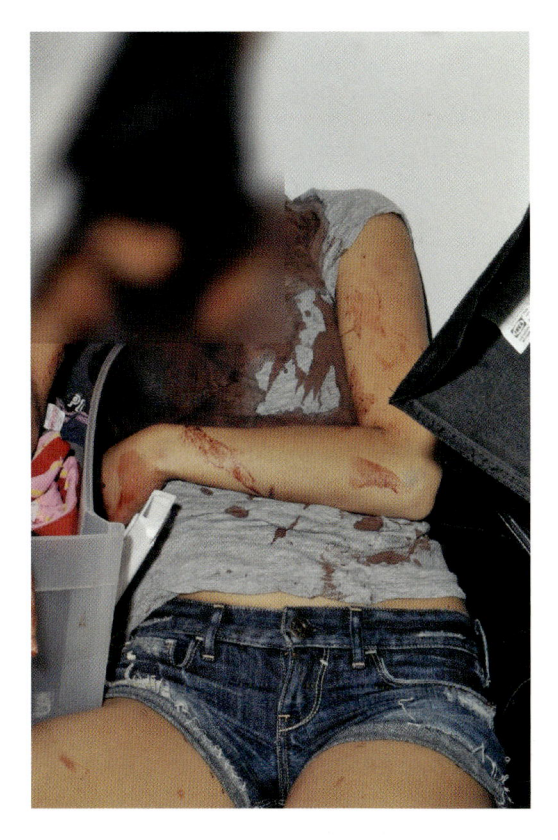

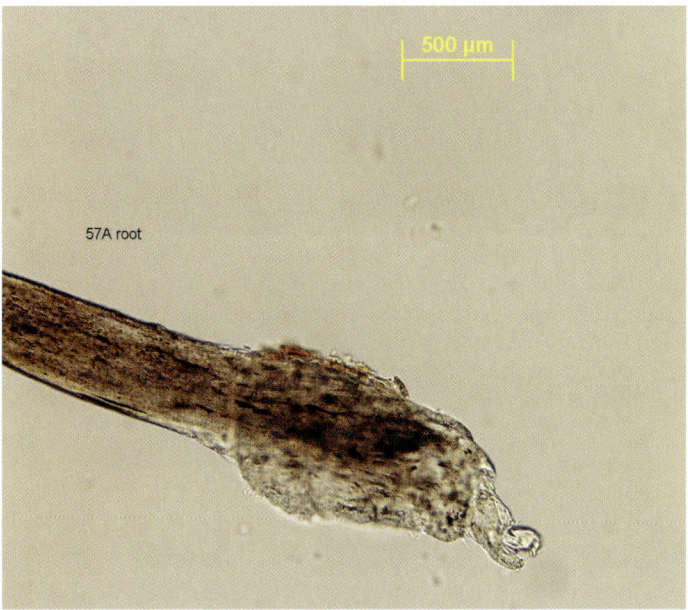

Figure 8.4 A female was found in her closet deceased with multiple stab wounds. Tape lifts were taken of her clothing. One lift taken from her back had a hair with a root present. DNA from the hair linked to a known individual as a possible contributor.

Figure 8.5 A male was found deceased in his residence with bindings on his wrists and ankles, and a blindfold on his face. The medical examiner listed the cause of death as undetermined and classified the death as a homicide. Swabs from the bindings revealed mixtures that linked the decedent and two suspects as possible contributors. (Article can be found at http://www.chron.com/default/article/Two-suspects-arrested-in-slaying-of-man-near-Gray-1717723.php.) (continued)

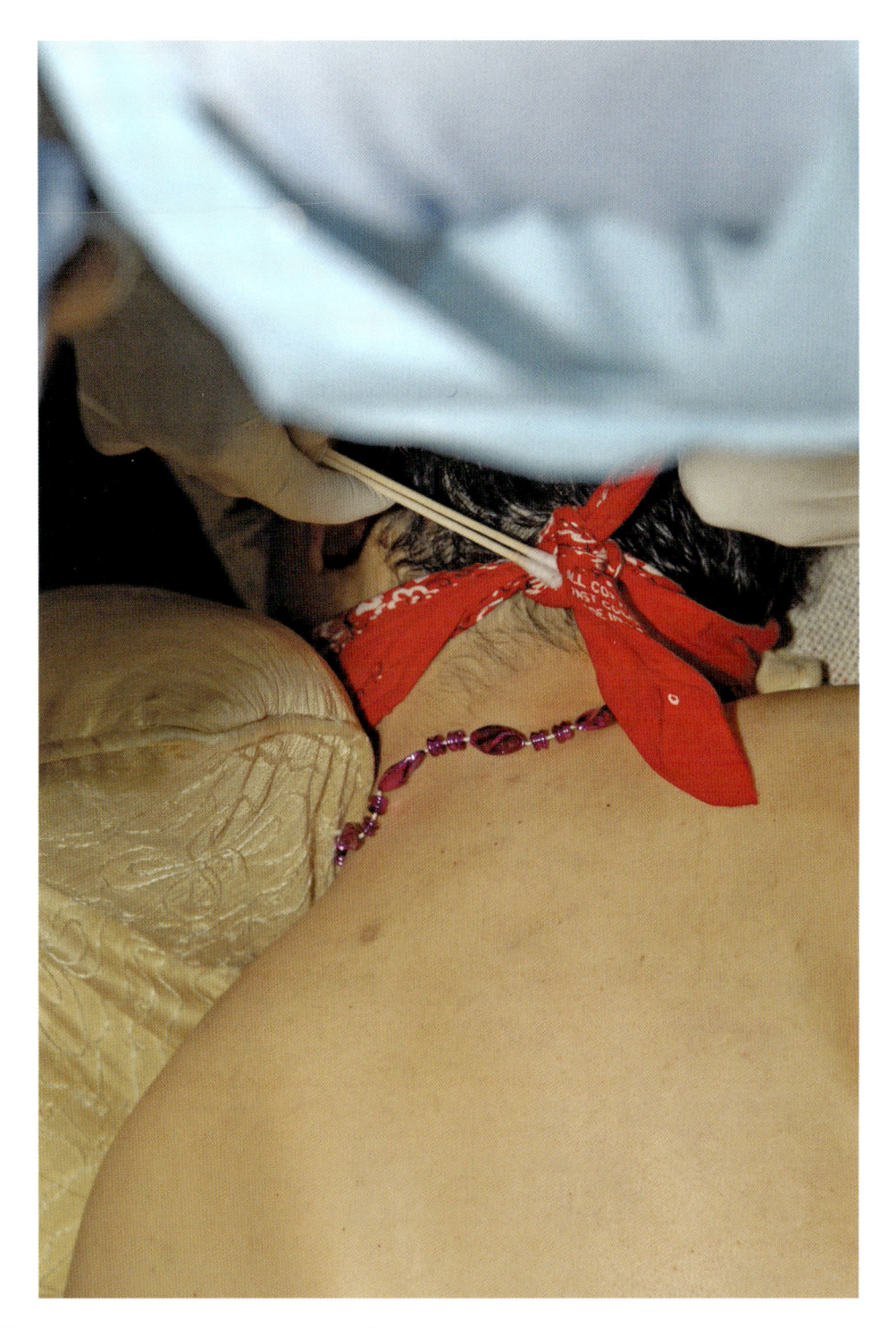

Figure 8.5 (continued) A male was found deceased in his residence with bindings on his wrists and ankles, and a blindfold on his face. The medical examiner listed the cause of death as undetermined and classified the death as a homicide. Swabs from the bindings revealed mixtures that linked the decedent and two suspects as possible contributors. (Article can be found at http://www.chron.com/default/article/Two-suspects-arrested-in-slaying-of-man-near-Gray-1717723.php.) (continued)

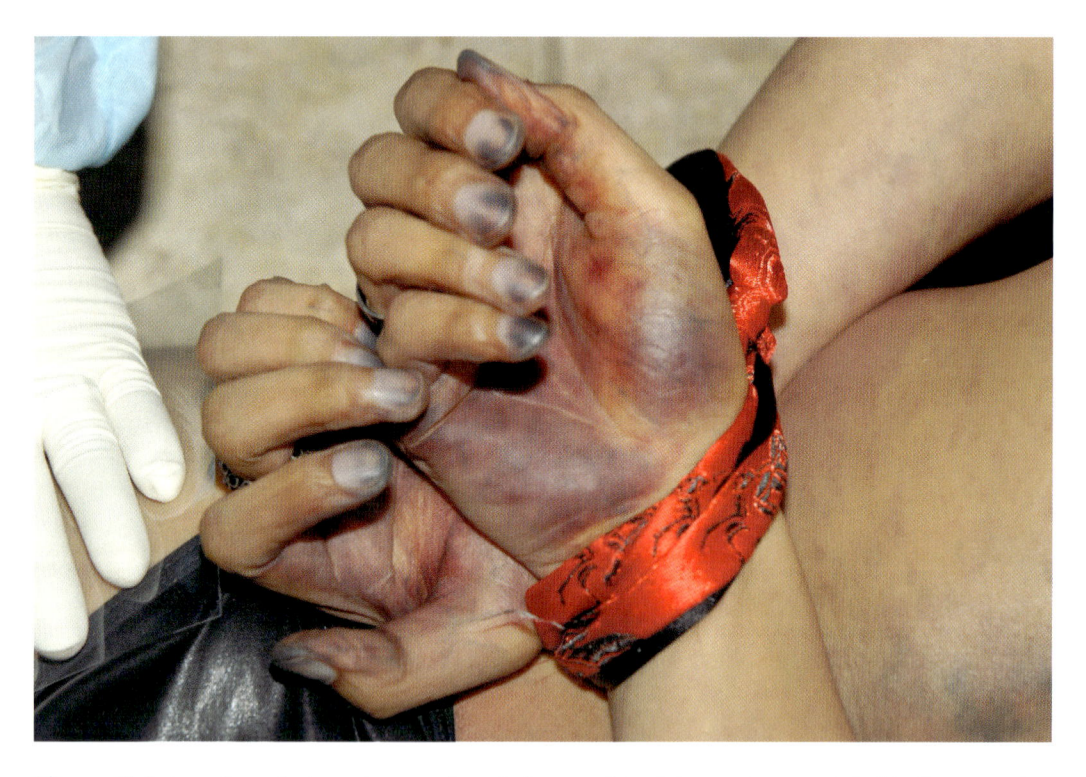

Figure 8.5 (continued) A male was found deceased in his residence with bindings on his wrists and ankles, and a blindfold on his face. The medical examiner listed the cause of death as undetermined and classified the death as a homicide. Swabs from the bindings revealed mixtures that linked the decedent and two suspects as possible contributors. (Article can be found at http://www.chron.com/default/article/Two-suspects-arrested-in-slaying-of-man-near-Gray-1717723.php.)

Figure 8.6 A nude female was found in the woods deceased from a gunshot wound. Handcuffs were used to hogtie her wrists and ankles. The trace analyst swabbed the handcuffs at the scene, but the collection was made difficult by her location in a ditch. After collection, the handcuffs were bagged and submitted to the laboratory for additional collection. A DNA mixture resulted that linked the decedent and a suspect as possible contributors.

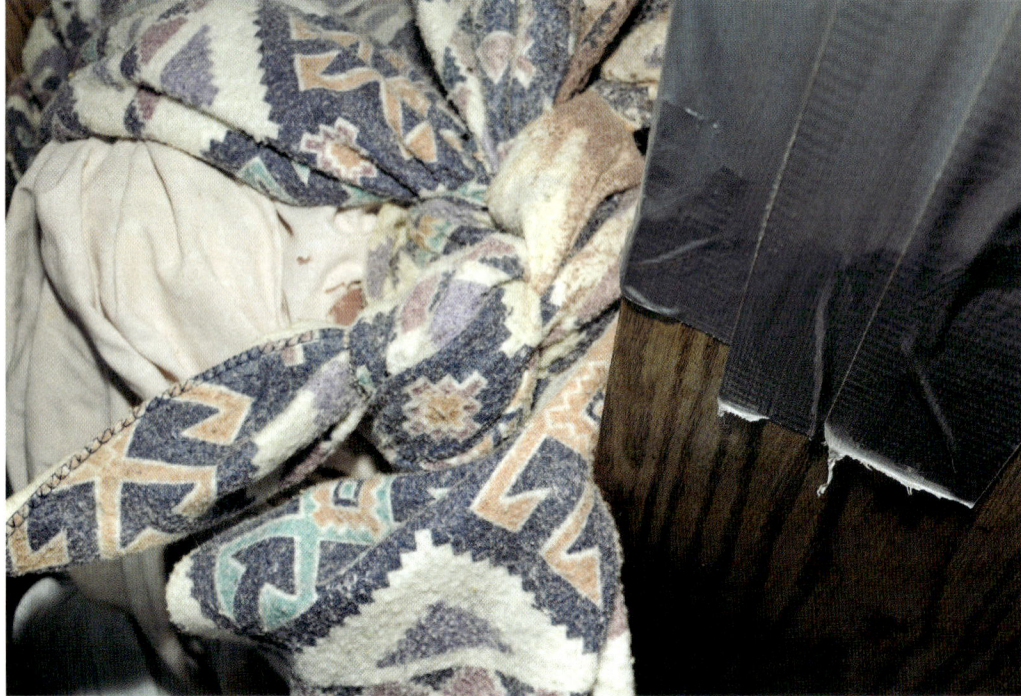

Figure 8.7 A male was found deceased in his closet, wrapped in a comforter. A sample from a swabbed knot in the comforter yielded a DNA mixture that linked the decedent and one of the suspects as possible contributors.

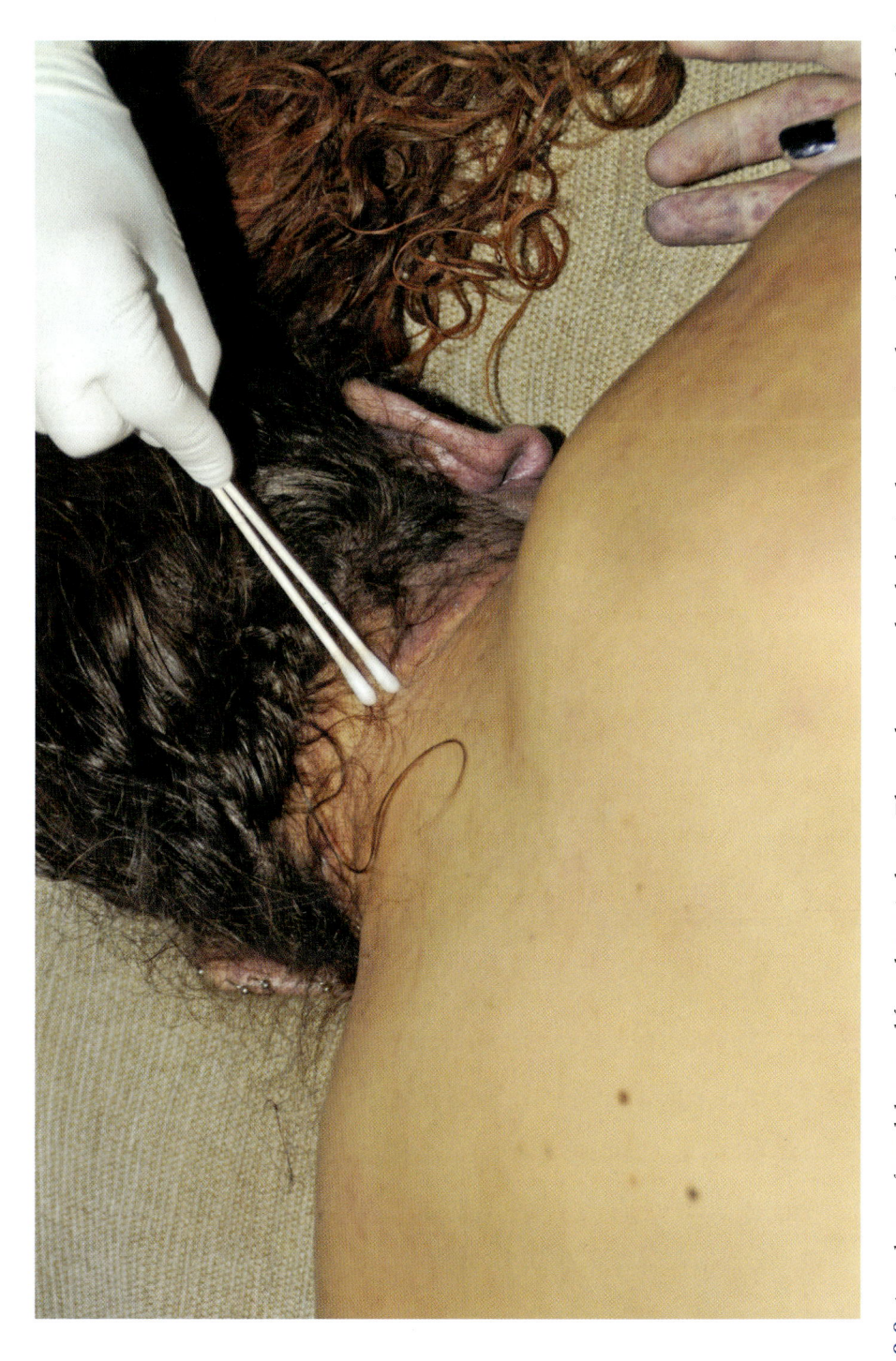

Figure 8.8 A male was found deceased face down indoors. The neck appeared to be bruised, suggesting he might have been strangled. Swabbings taken from the neck revealed a DNA mixture that linked the decedent and a suspect as possible contributors. The pathologist collected samples from under the decedent's fingernails, which also produced a mixture that linked the decedent and the suspect as possible contributors. The decedent's hands were protected at the crime scene with paper bags to guard against loss of foreign DNA and gunshot residue.

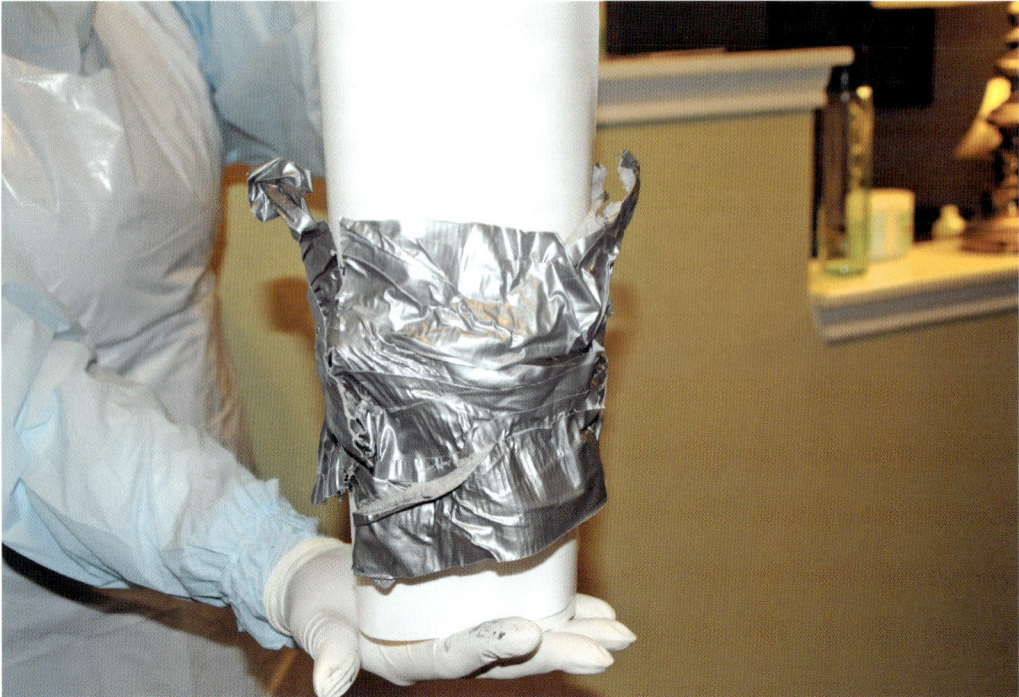

Figure 8.9 A male decedent with a gunshot wound was found in a car trunk bound with duct tape. The tape, which was not in proximity to the wound, was cut off of the body and preserved. The sticky side was placed on the shiny side of laboratory bench paper to preserve fingerprints and DNA. DNA testing of the tape revealed a mixture linking the decedent and at least one more unidentified individual as possible contributors. (continued)

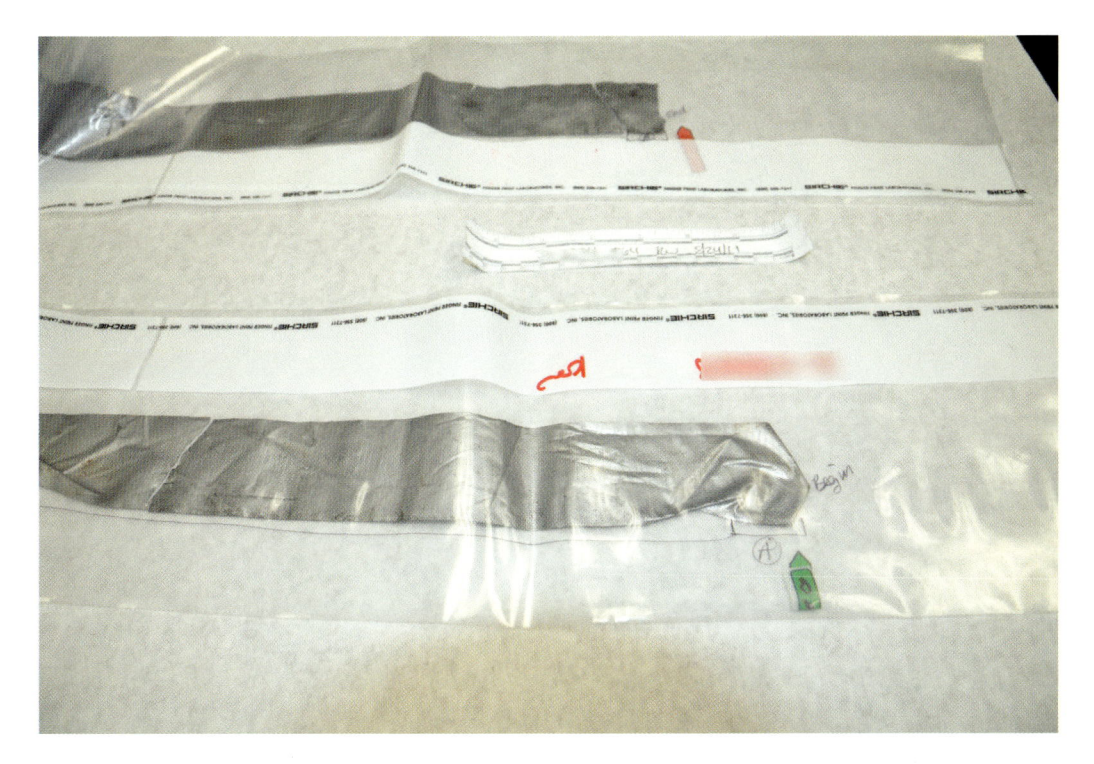

Figure 8.9 (continued) A male decedent with a gunshot wound was found in a car trunk bound with duct tape. The tape, which was not in proximity to the wound, was cut off of the body and preserved. The sticky side was placed on the shiny side of laboratory bench paper to preserve fingerprints and DNA. DNA testing of the tape revealed a mixture linking the decedent and at least one more unidentified individual as possible contributors.

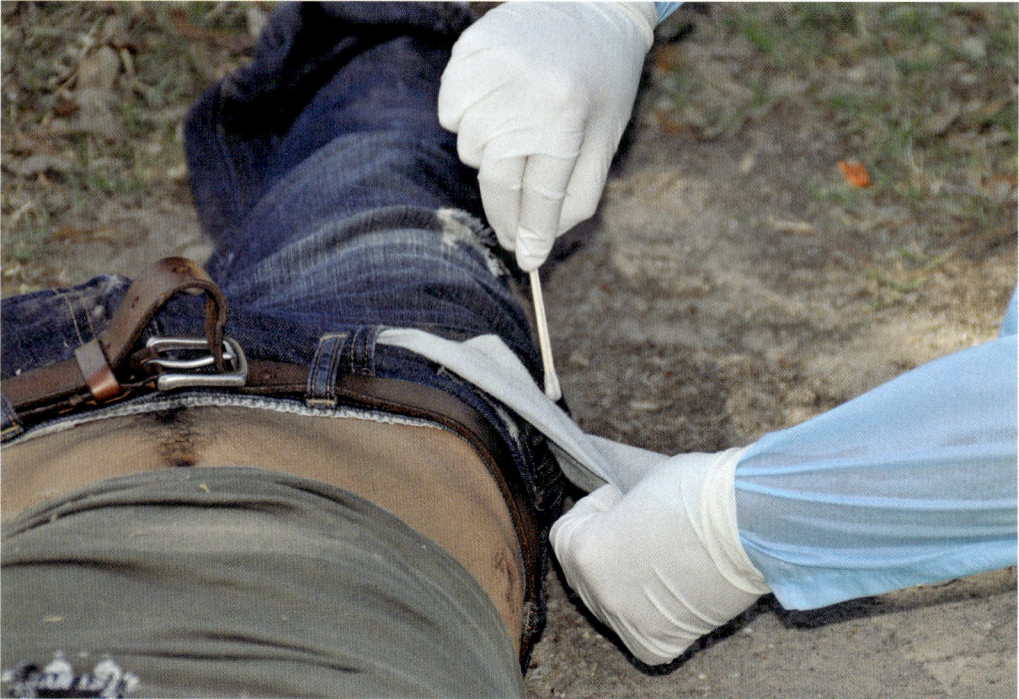

Figure 8.10 A male decedent was found face up in a street with a gunshot wound. The pockets on his pants were turned out, suggesting activity by someone other than the decedent. The swabbing from the pockets revealed a mixture that linked the decedent and at least one unknown individual as possible contributors.

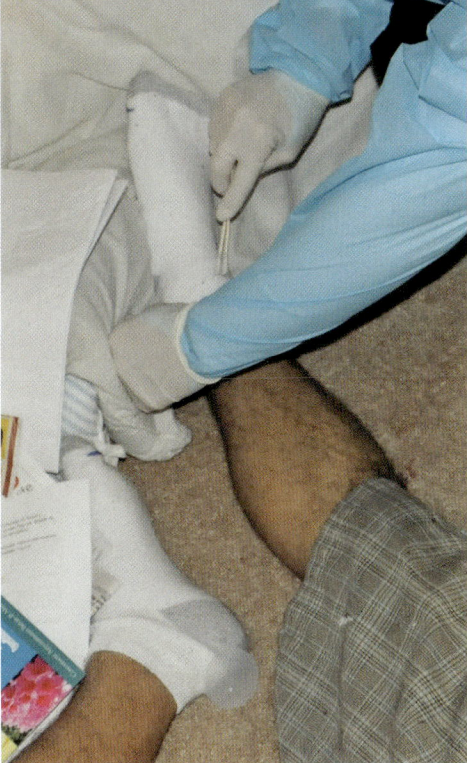

Figure 8.11 A male decedent was found face up in his residence with multiple sharp force injuries. Marks on the carpet suggested he was dragged, possibly by his ankles. The body was tape lifted and swabbed, including the ankles. A mixture of DNA was found on the ankles that linked the decedent and a suspect as possible contributors.

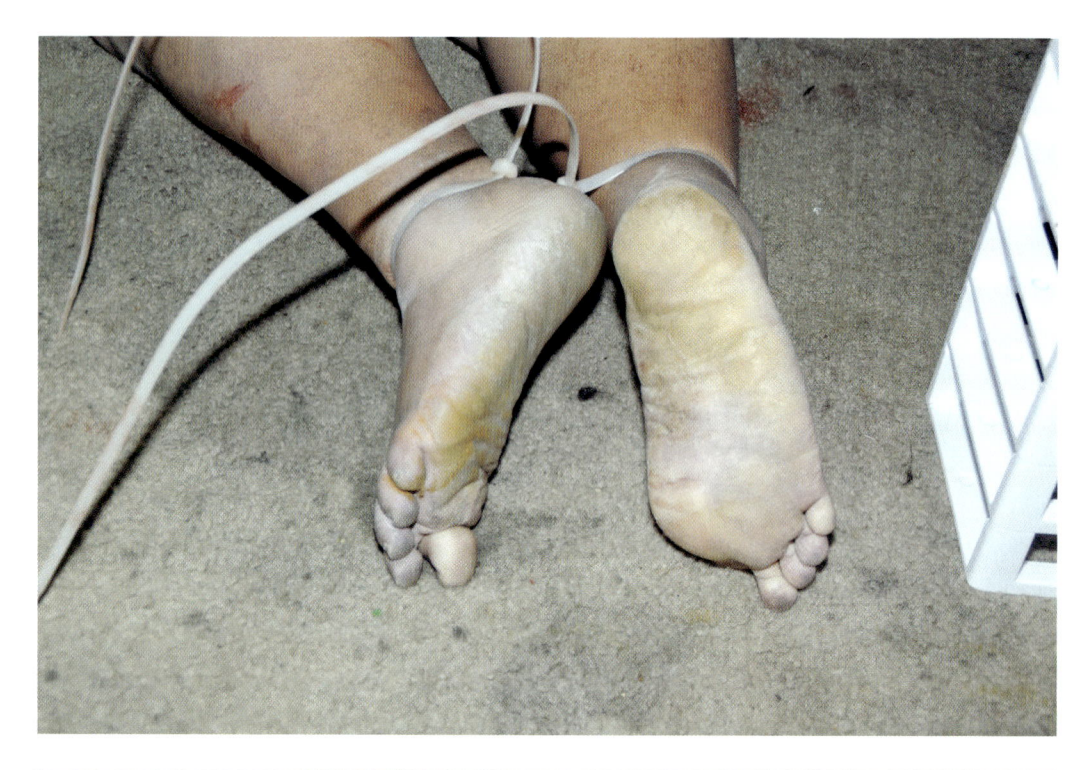

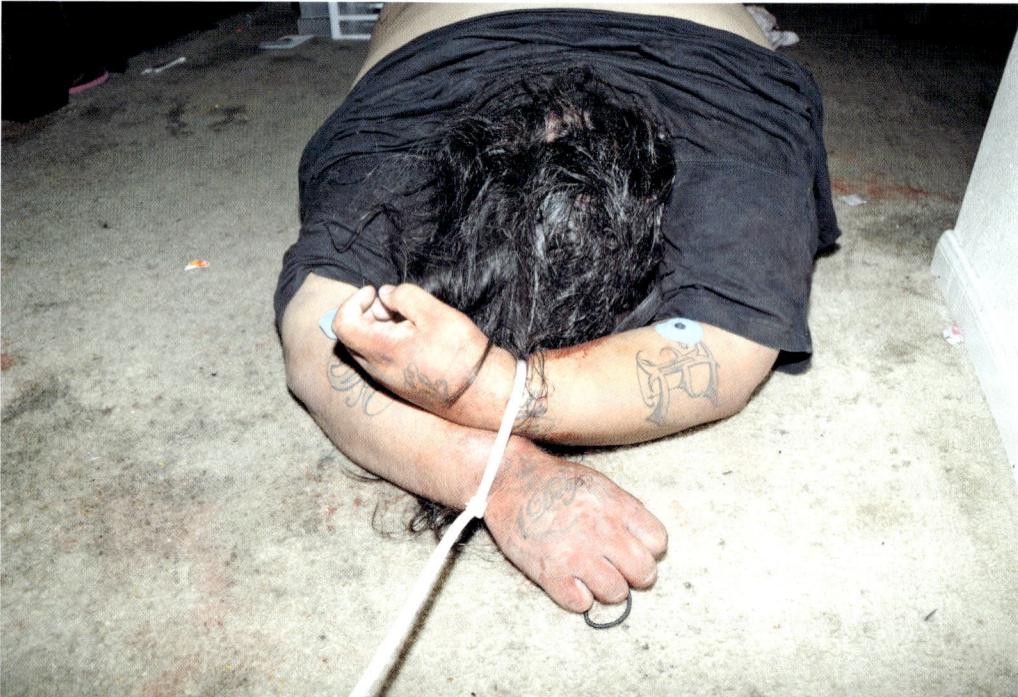

Figure 8.12 A male decedent was nude from the waist down, face down, on the ground. He was bound with zip ties on his ankles and wrists, which were swabbed as part of the trace collection. The DNA from the zip ties revealed a mixture that linked the decedent plus a small number of alleles not from the decedent.

Trace Success Cases 143

Figure 8.13 A female decedent was found nude, face down, near a cemetery. There were no visible wounds, but there were blood swipes on the back of her neck, buttocks, calf, and on the bottom of her foot. The blood swipes were swabbed as part of the collection. The swipes revealed a mixture of DNA that linked the decedent and at least one unknown male. A profile from the swipe of the calf led to a CODIS match to a suspect. (Article can be found at http://www.chron.com/news/houston-texas/article/Houston-police-make-arrest-in-March-slaying-of-2683082.php.) (continued)

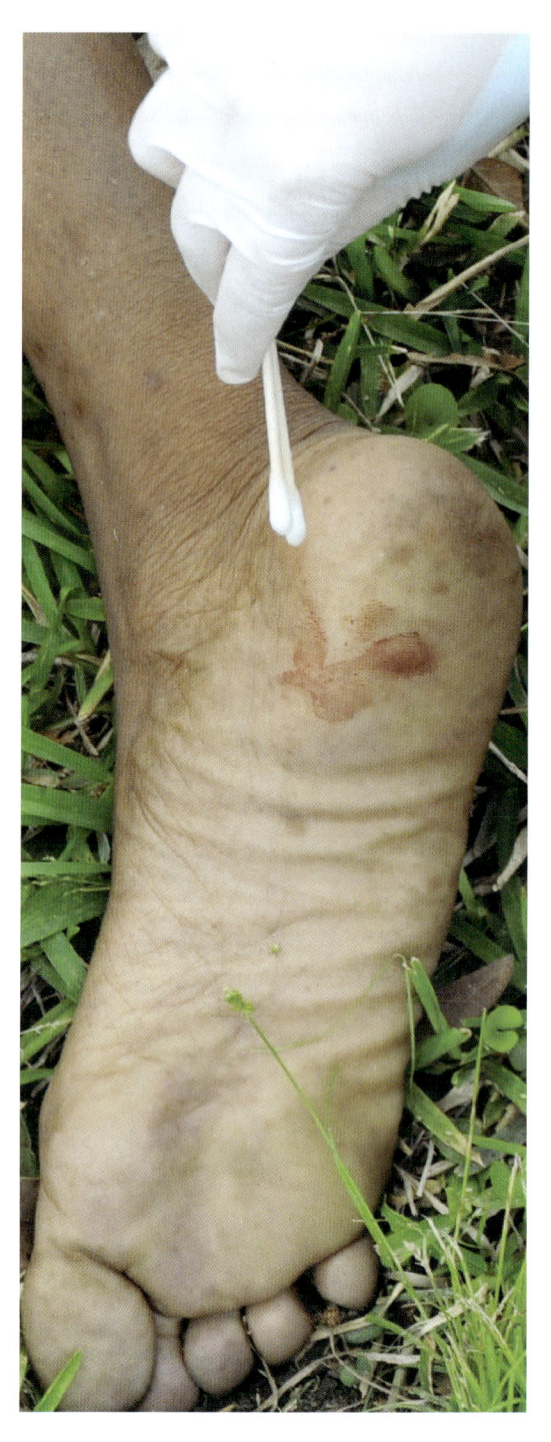

Figure 8.13 (continued) A female decedent was found nude, face down, near a cemetery. There were no visible wounds, but there were blood swipes on the back of her neck, buttocks, calf, and on the bottom of her foot. The blood swipes were swabbed as part of the collection. The swipes revealed a mixture of DNA that linked the decedent and at least one unknown male. A profile from the swipe of the calf led to a CODIS match to a suspect. (Article can be found at http://www.chron.com/news/houston-texas/article/Houston-police-make-arrest-in-March-slaying-of-2683082.php.)

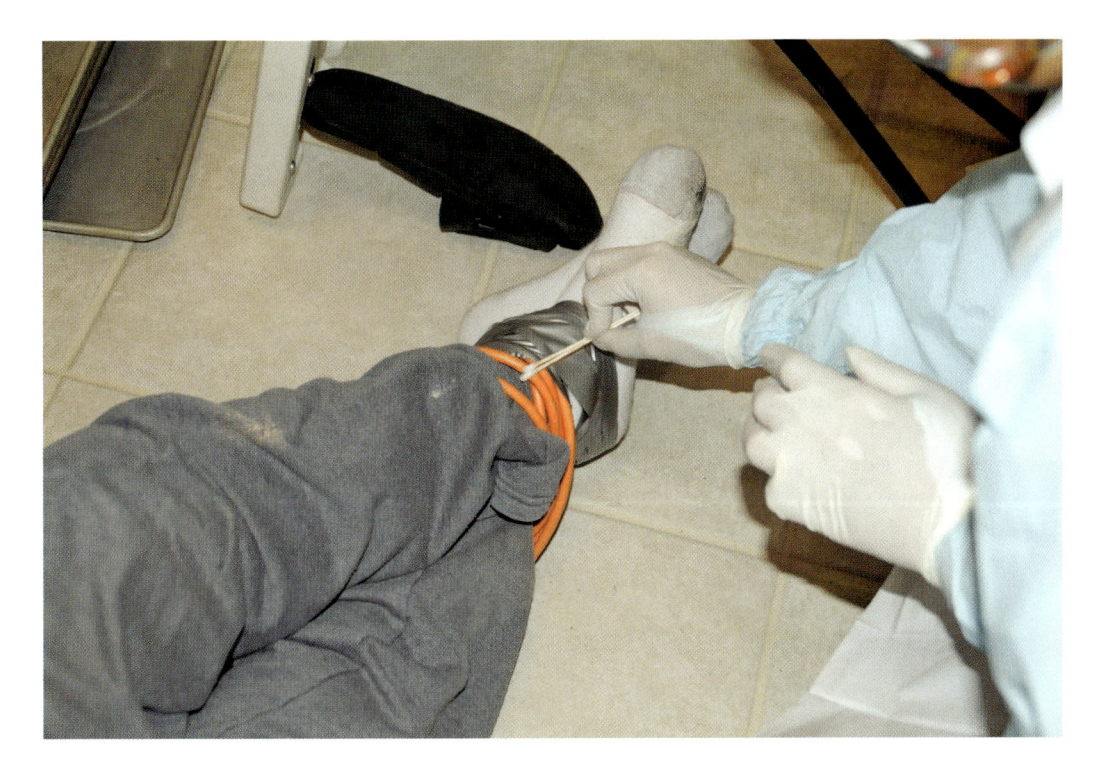

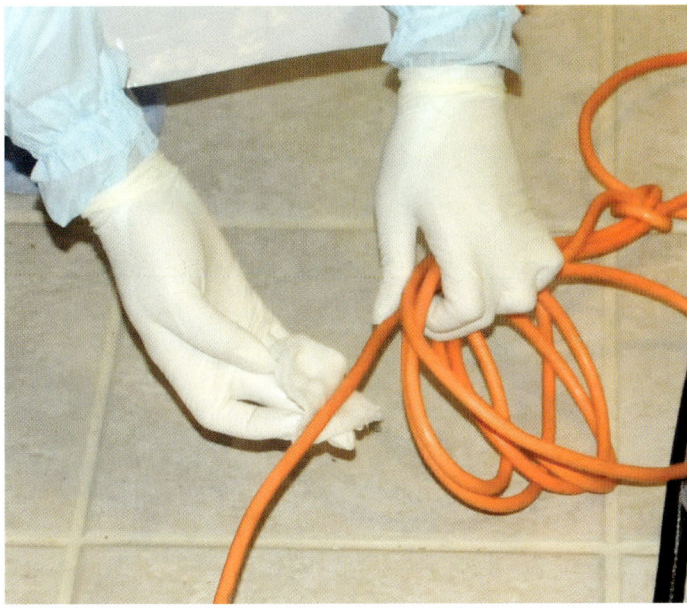

Figure 8.14 A male decedent was found in his house, electrical cords and duct tape wrapped around his wrists and ankles. Because of the long length of the electrical cord, a gauze pad was used instead of a swab or a series of swabs. The cord swabbings linked to a suspect as a possible contributor.

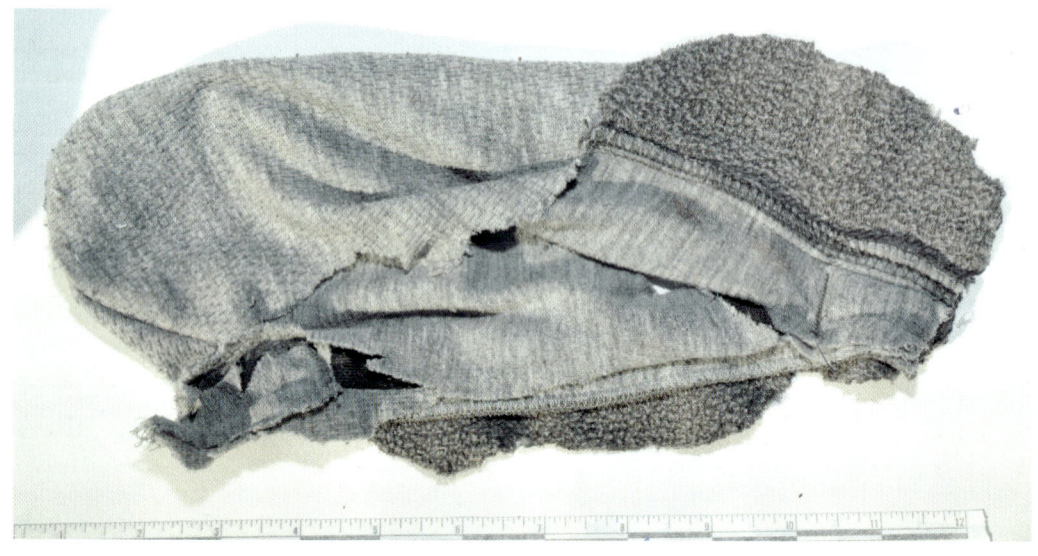

Figure 8.15 A male decedent was found face down in the grass with a gunshot wound. In this case, once again, a pocket was turned out. It was cut from the pants and preserved. The pocket revealed a DNA mixture that linked a suspect as a possible contributor.

Trace Success Cases 147

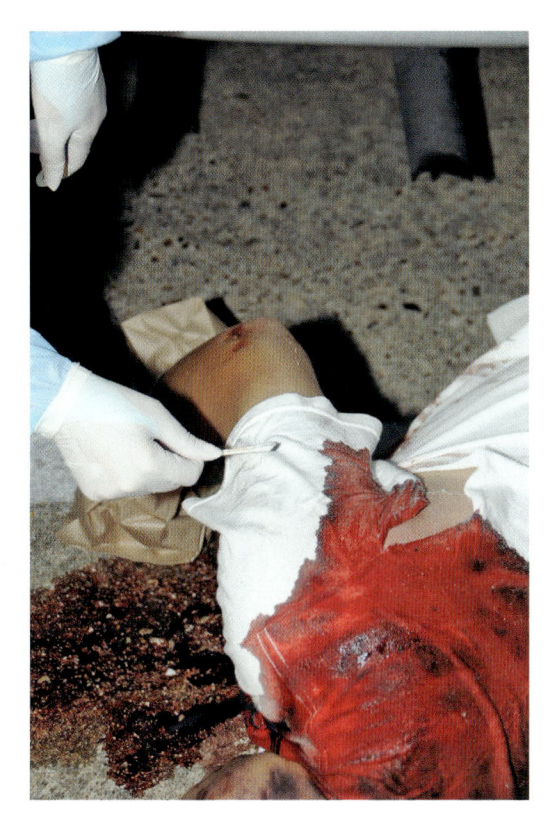

Figure 8.16 A male decedent was found face up in a parking lot with a gunshot wound and multiple stab wounds. Swabs were collected at the shoulder away from the bloody area as well as from the hands and a turned-out pocket. The shoulder swabs produced a mixture that linked the decedent as a possible contributor with a few additional alleles.

Figure 8.17 A male decedent was found on his side in a grass field with a gunshot wound. Since his right hand was bloody, samples were not collected. Swabs of the knuckles of his left hand produced a DNA mixture that linked the decedent plus several additional alleles from one or more unknown contributors.

Trace Success Cases

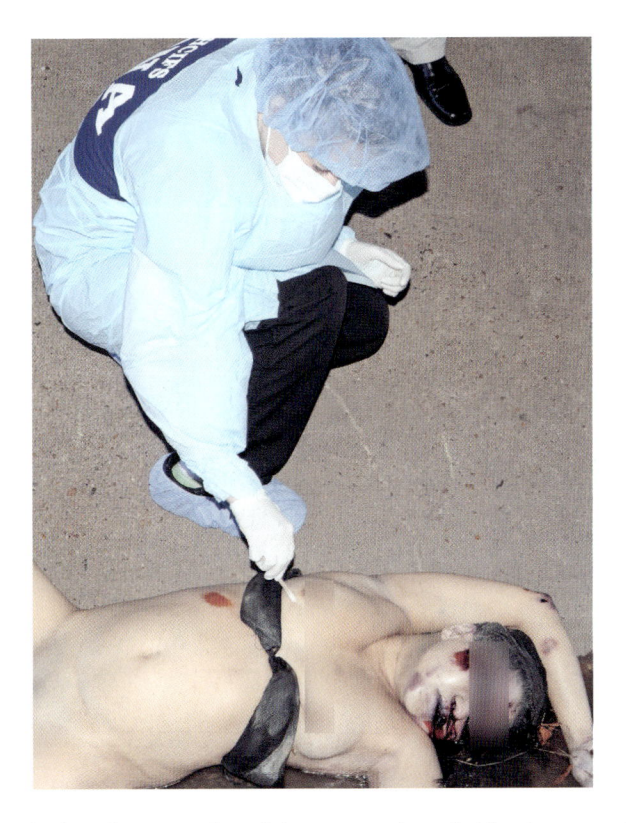

Figure 8.18 A female decedent was found face up under a bridge in water in a ditch wearing only a bra below her breasts. She had bruises and scrapes in several locations. Swabbings of her breasts and bruises, and tape lifts of her body were collected. DNA from the breast swabs and from hairs on the tape lifts linked to a suspect as a potential source.

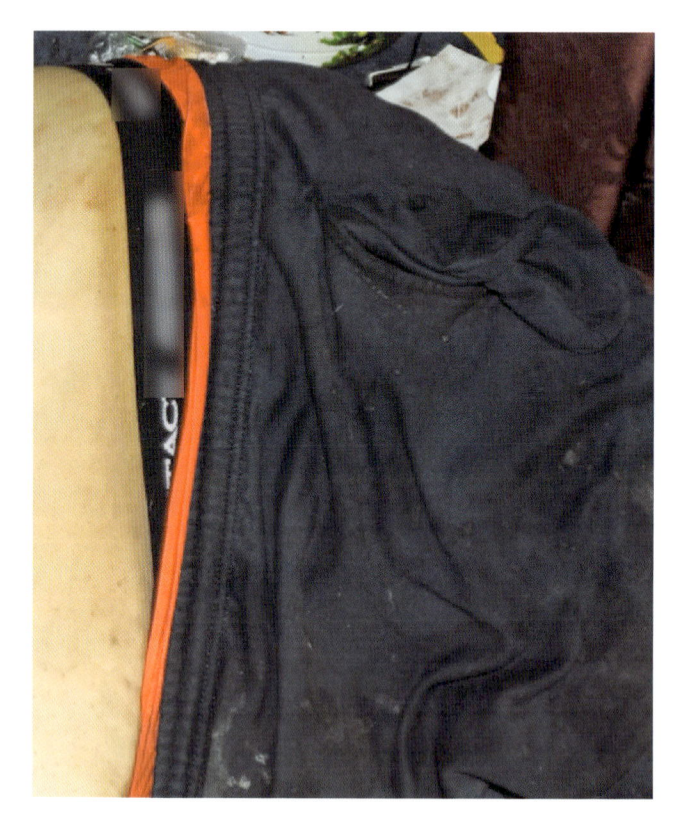

Figure 8.19 A deceased male was found in his apartment on his side with one of his pockets turned out. DNA from swabs of the pocket produced a mixture that linked the decedent and a suspect as possible contributors.

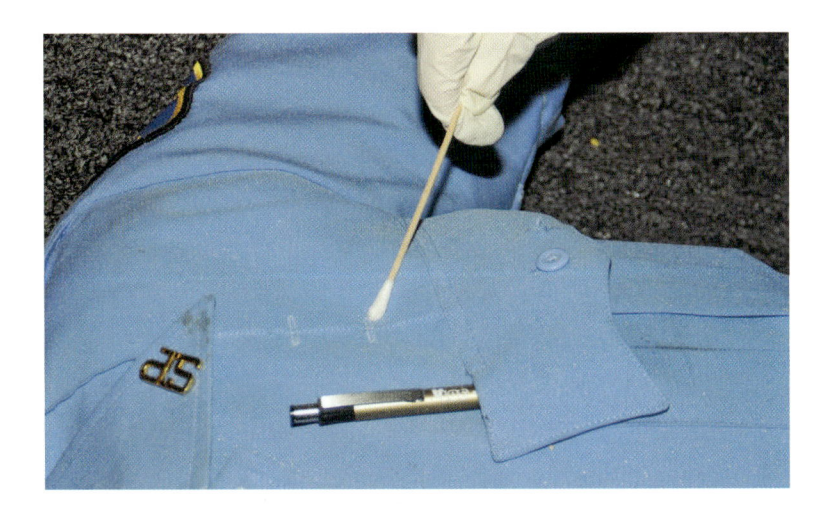

Figure 8.20 A deceased male security guard was found face up in an illegal gambling hall with a gunshot wound of the head. The guard's badge had been removed from his shirt and his weapon was removed from its holster. Swabbings of the shirt from the badge area revealed a DNA mixture that linked the decedent and an unknown individual as possible contributors.

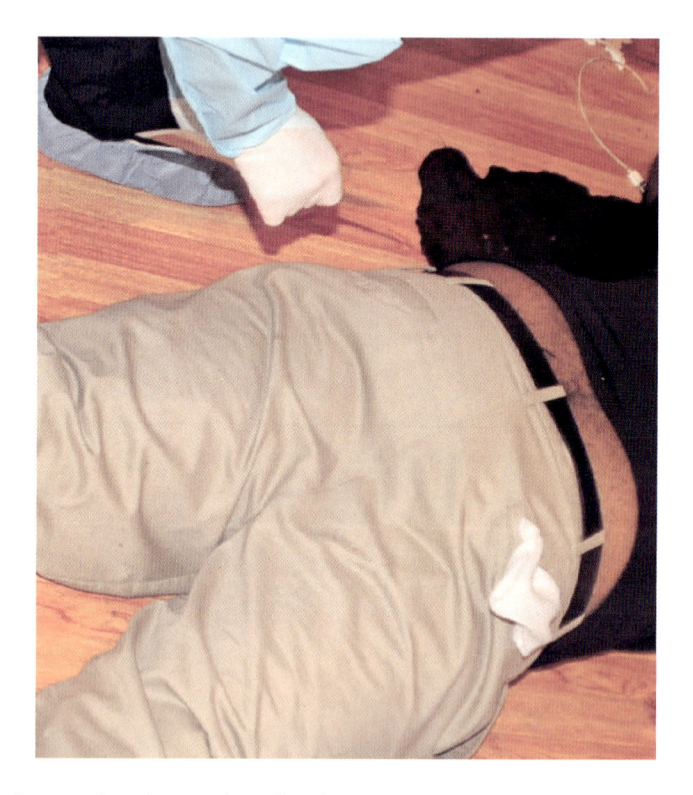

Figure 8.21 A deceased male was found in his apartment with a gunshot wound. Swabbings of his turned-out back pocket revealed a mixture that linked the decedent and a suspect as possible contributors.

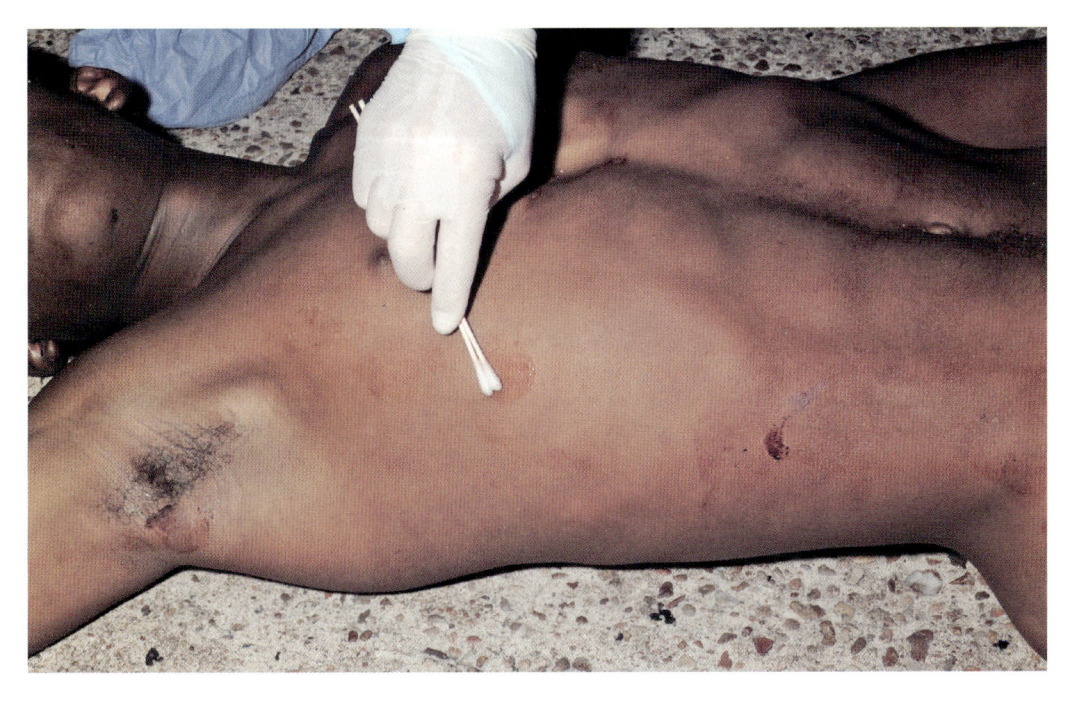

Figure 8.22 A deceased male was found face up in the street with gunshot wound. The suspect was wearing pants and shoes, but no shirt. Swabbings were taken of his hands, neck, armpits, ankles, and a clear fluid stain on his chest. The chest stain revealed a DNA mixture that linked the decedent and an unknown individual as possible contributors.

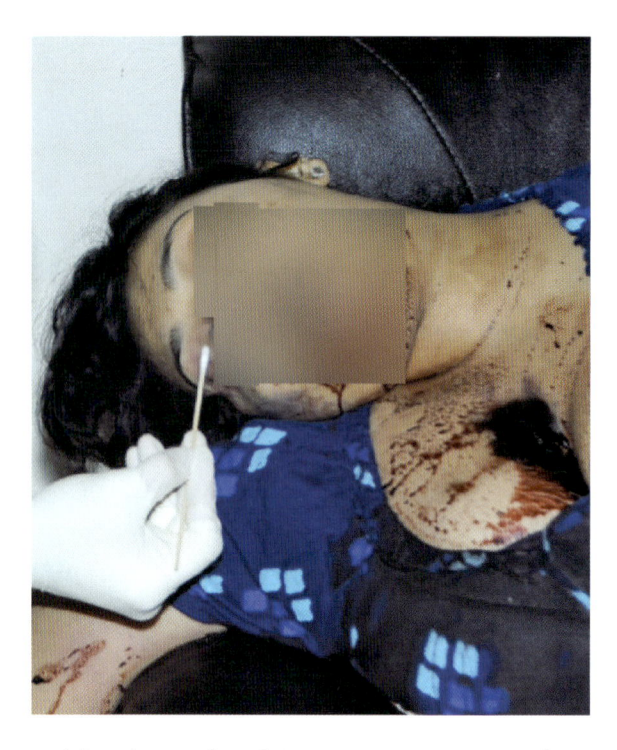

Figure 8.23 A deceased female was found in a sitting position on the couch in her residence with a gunshot wound of the chest and a bruise above her right eye. Swabbing of her hands and the bruise above her eye were taken. The sample from the bruise returned a DNA mixture linking one or more unknown individuals as possible contributors.

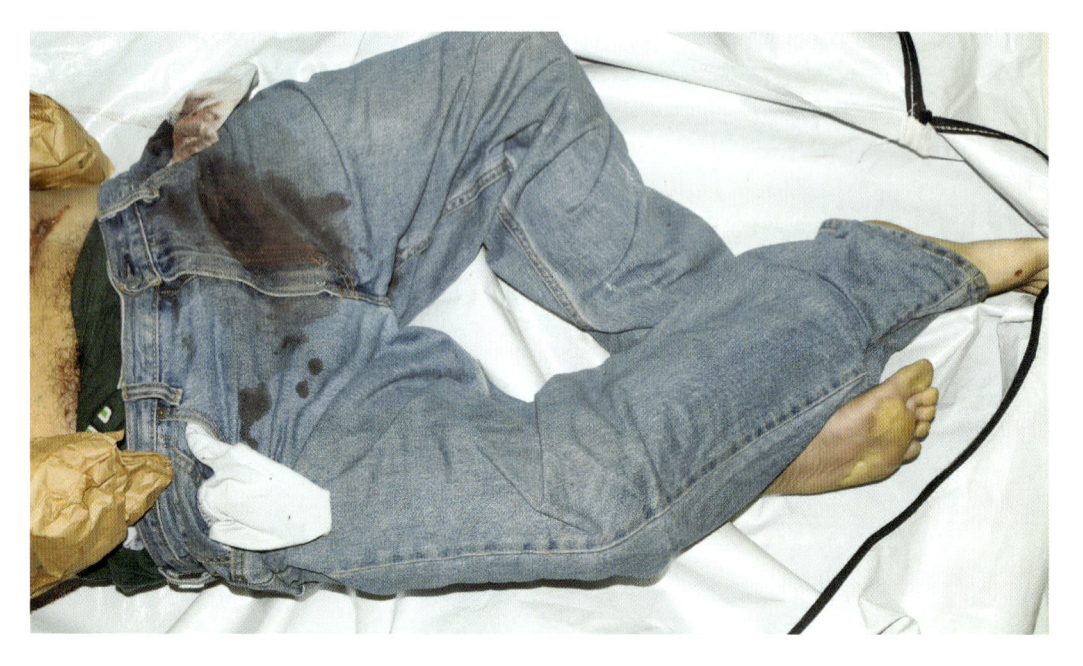

Figure 8.24 A deceased male was found outdoors with pants pockets turned out. Trace collection was performed in the morgue, not at the scene. Swabbing of the right pocket resulted in a DNA mixture that linked the decedent and several alleles from an unknown individual.

Figure 8.25 A deceased female was found inside her garage wrapped in a rug. Multiple stab wounds to her body were observed. The rug was swabbed (and taped) at the scene and then placed in a separate body bag to preserve bloodstains. The rug was also swabbed in the laboratory, and one stain revealed a DNA mixture that linked the decedent and another individual as possible contributors. (continued)

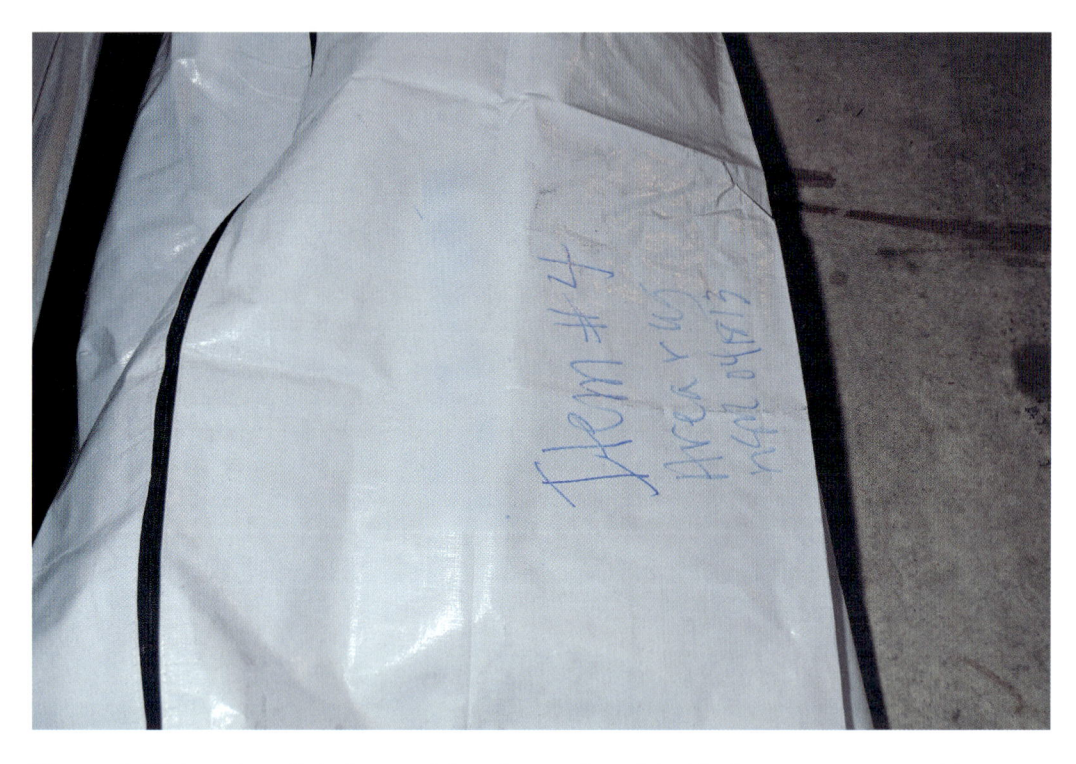

Figure 8.25 (continued) A deceased female was found inside her garage wrapped in a rug. Multiple stab wounds to her body were observed. The rug was swabbed (and taped) at the scene and then placed in a separate body bag to preserve bloodstains. The rug was also swabbed in the laboratory, and one stain revealed a DNA mixture that linked the decedent and another individual as possible contributors.

Administration

9

Developing a trace team can be a challenge. Trace DNA samples are often collected from apparently unstained surfaces such as turned-out pockets and bindings. Qualified DNA analysts are well equipped to collect such samples from a decedent. But DNA analysts are not the only ones trained and experienced in sterile techniques and familiar with the remarkable sensitivity of forensic DNA tests. For agencies without an onsite DNA laboratory, decedent DNA evidence can be collected by trained medical examiner's investigators and pathologists (at the scene or in the morgue) or by trained and qualified crime scene personnel if given access to a decedent at a scene. Proper and accurate documentation of the collection must be enforced. If the collection is performed by non-ME personnel, the documentation of the collection (including photographs) must be available to the pathologist prior to the postmortem examination. Regardless of who collects samples from a decedent, contamination must be a constant concern. Training is essential to ensure effective and comprehensive trace DNA evidence collection.

Standard operating procedures must be developed for trace DNA evidence collection activities. The procedures must be written to accommodate a variety of scenes. Each trace scene is unique, and analysts must be imaginative to be effective. Collection techniques should be fully described, and proper labeling and packaging procedures should be included. Policies should be developed in conjunction with the forensic pathology staff so as not to interfere with their work. Formal external review of trace DNA evidence collection procedures and quality assurance via accreditation are strongly recommended. Crime scene accreditation is available from the American Society of Crime Laboratory Directors Laboratory Accreditation Board (ASCLD/LAB)[*] and from Forensic Quality Services (FQS).[†]

For the TECT to function properly, a callout protocol must be put in place to define the circumstances that initiate calling the TECT to a crime scene. We recommend an advisory board to review and guide TECT policy. As described above, the HCIFS Advisory Board includes the trace team lead, several forensic pathologists, death investigators, a trace evidence lab manager, a DNA lab manager, an anthropologist, and the crime lab director.

The board reviews lessons learned from problems that arise and, at times, implements policy modifications. For example, the board met to consider ways to reduce the number of callouts of TECT analysts to crime scenes. For the past year, TECT analysts have attended shooting death scenes if there was evidence of a struggle or other contact between the assailant and decedent before or after the shooting. Surprisingly, informative DNA results from shootings were successful as frequently as those from bound or dumped decedents. The board developed a procedure to return most shooting victims to the morgue without on-scene trace collection, mindful of the need to limit disruption of body fluids during

[*] www.ascld.org
[†] fqsforensics.org

transport. This policy preserved trace evidence while reducing the number of callouts by volunteer TECT analysts.

The nine-analyst HCIFS trace team is headed by a coordinating manager mindful of the needs of the pathologists, the laboratory, and the investigating agencies. A balance must be maintained between the trace team and the investigating agencies. TECT analysts are at the crime scene to collect evidence for the investigating law enforcement agency. But frequently the TECT collection extends time that law enforcement investigators spend at a scene. The manager is in constant contact with agency supervisors to ensure proper communication. Successes of the trace team are shared with the investigating law enforcement agency to reinforce the value of the evidence TECT analysts collect.

Closing Remarks

As we traveled along this path, we did not anticipate the successes we encountered. As the story has unfolded, we have found surprising locations where foreign DNA has been hiding. Personal objects that an individual touches every day (pockets, belts, etc.) are frequently found to contain high amounts of foreign DNA from suspects after one encounter. Suspect DNA can be collected directly off of the skin of a decedent and not be diluted out by the decedent's DNA. This information opens another door in the field of forensics and allows us to collect DNA from locations that seemed, in the past, not to be worthwhile. We hope that every agency takes a walk down this path and finds the successes we have found during this journey.

References

Graham, E., and Rutty, G. 2008. Investigation into "normal" background DNA on adult necks: Implications for DNA profiling of manual strangulation victims. *J For Sci* 53(5):1074–1082.
Kenna, J., Smyth, M., McKenna, L., Dockery, C., and McDermott, S. 2011. The recovery and persistence of salivary DNA on human skin. *J For Sci* 56(1):170–175.
Kirk, P.L. 1953. *Crime investigation*, 4. New York: Interscience Publishers.
Locard, E. 1920. *L' enquete criminelle et les methodes scientifiques*, 139. Paris: Flammarion.
May, R., and Thomson, J. 2009. Optimisation of cellular DNA recovery from tape-lifts. *For Sci Intl Genet* 2(1):191–192.
Moser, M.R. 2013. Pocket DNA conviction under appeal: Bruff opts to defend self in series of hearings. http://crossville-chronicle.com/local/x2000924396/Pocket-DNA-conviction-under-appeal.
Saferstein, R. 2009. *Forensic science: From the crime scene to the crime lab*. Upper Saddle River, NJ: Prentice Hall.
SWGSTAIN terminology. 2009. *Forensic Science Communications*, April, 11(2).
Texas Code of Criminal Procedure (CCP). Article 49.25.
Williamson, A.L. 2012. Touch DNA: Forensic collection and application to investigations. *J Assoc Crime Scene Reconstr* 18(1):1–5.

Index

A

Accreditation, 157
Administration, 157–158
Alternate light sources, 18, 19f, 118f
American Society of Crime Laboratory Directors Laboratory Accreditation Board (ASCLD/LAB), 157
Amido black staining, 82f, 83f
Ankles, 53, 62–63f, 81f, 141f
Apron, 10, 18f
Armpits, 53
Arms, swabbing, 64f

B

Belly button, 53
Belly button rings, 65f
Belts, 53, 92f
    used as bindings, 96f
    used for strangulation, 97f
Bindings, 2, 50f, 85, 96f, *See also* Duct tape evidence
    belts as, 96f
    bullet holes in, 37
    cloth restraints, 102f
    gauze swabbing, 109f
    handcuffs, 108f, 135f
    multiple types, 101f
    shoelaces as, 100f
    successful trace analyses, 132–135f, 142f, 145f
    zip ties, 106–107f, 142f
Blindfolds, 132f
Blood patterns, 38, 50f, 73
    amido black staining, 82f, 83f
    on clothing, 81f
    evidence of dragging, 81f
    ninety degree blood drops, 73, 74f, 75f, 76f
    on shoes or feet, 75f, 76f, 77f, 144f
    smears, 79f, 80f
    swipes, 73, 78f, 143–144f
    void patterns, 53, 71f, 72f, 73, 87f
Bodies
    approaching at crime scene, 53
    documenting interventions with, 9, 26f
    evidence of dragging, 81f, 141f
    knife in, 37, 38f, 39f
    moving and handling, 11f, 37
    preserving/packaging hands, 38, 47f, 49f, 50f
    swabbing, *See* Swabbing
    TECT worksheet, 15f
Body jewelry, 53, 65f
Body swabbing, *See* Swabbing
Body wrappings, *See* Wrappings
Breast swabbings, 53, 61f, 149f
Bridges, bodies under, 125f, 149f
Bruises, swabbing, 53, 69–70f, 154f
Bullet holes, 37, 128f
Buttons, 92f

C

Call-out criteria, 2, 157–158
Car trunks, 120f, 121f, 138–139f
Chain of custody, 14f, *See also* Documentation
Cloth bindings, 102f
Clothing, 85
    blood patterns on, 81f
    bullet holes in, 37
    murdered officer's, 86f, 151f
    pants belts and buttons, 92f
    pockets, 38, 47f, 48f, 85, 89–91f, 140f, 150f, 152f, 154f
    preservation, 38, 48–50f
    tape lifts, 94f, 130–131f
CODIS (Combined DNA Index System), 1, 85, 127
Comforter, as body wrapping, 113f, 136f
Condoms, 67f
Continuing education, 5
Cord bindings, 85, 101f, 103–105f
    gauze swabbing, 109f, 145f
Crime scene accreditation, 157
Crime scenes
    approaching the decedent, 53
    call-out criteria for trace evidence collection team, 2, 157–158
    safety considerations, 122f
    TECT analyst interactions at, 6–7
    TECT supplies and materials, 7–9, 12f
Crime scenes, tough places, 119
    under bridge, 125f
    car trunks, 120f, 121f, 138–139f
    ditches or water, 124f
    loaded weapons, 119, 120f
    safety considerations, 122f
    tight areas, 119, 121f, 123f, 126f

D

Debris tape lifts, 10
Deionized water, 22f
Dismembered bodies, 71f
Ditches, 124f
    under bridge, 125f, 149f
DNA recovery, *See* Evidence collection techniques; Touch DNA; Trace evidence analysis, successful DNA recovery
DNA transfer, unrelated to investigation, 53
Documentation, 9, 157, *See also* Labeling evidence
    evidence cut from decedent, 51f
    interventions with bodies, 9, 26f
    TECT submission form, 14f
    TECT worksheet, 15f
Drying wet items, 38
Duct tape evidence, 37, 41f, 42f, 85, 98–99f, 101f, 128f, 138–139f

E

Electrical cords, 85, 101f, 103–105f, 145f
Emergency medical technician interventions, 9
Epithelial cells, 14, 28f, 85, *See also* Touch DNA
Error correction, for evidence labels, 44f
Evidence collection documentation, *See* Documentation
Evidence collection techniques, 10, *See also* Swabbing; Tape lifts
    alternate light sources, 18, 19f, 118f
    picking using tweezers, 10, 29–32f
    standard operating procedures, 157
    wet bodies and, 29–30f
Evidence preservation, *See* Preservation
Exchange principle for evidence transfer, 5

F

Face swabbing, 59f, 70f
Feet, blood patterns on, 75f, 76f, 77f, 144f
Fiber evidence, 85, 93f
    tape lifts, 10, 28f
Fingernails, 49f, 137f
Fingerprint evidence
    amido black staining, 82f, 83f
    duct tape, 41f, 128f
    knife handle, 40f
Fingertips, 53, 54f, 55f
Firearms, loaded, 119, 120f
Forensic Quality Services (FQS), 157

G

Gauze swabbing, 18, 19, 34f, 109f, 145f
    labeling, 35f
Glove liners, 17f, 18, 19

Gloves
    analyst equipment, 10, 16f
    swabbing, 88f
Goggles, 18, 19f, 118f
Gowning, 10

H

Hair cover, 10
Hair evidence, 85, 131f
    tape lifts, 10, 28f
Handcuffs, 85, 108f, 135f
Hands, of decedent, 38, 47f, 49f, 50f
    swabbing knuckles or fingertips, 53, 54f, 55f, 148f
Harris County Institute of Forensic Sciences, 1
    TECT program, 1–2, 157–158, *See also* Trace DNA evidence collection team
Headlamps, 18, 19, 20f

I

Interventions with body, documenting, 9, 26f

J

Jewelry, 53, 65f, 68f

K

Knife
    in body, 37, 38f, 39f, 129f
    fingerprint preservation, 40f
Knots
    in body wrappings, 113–115f, 136f
    in cloth bindings, 102f
    in cords, 103–105f
    in shoelaces, 100f
Knuckles, 53, 54f, 148f

L

Labeling evidence, 32f, 34f, 35f, 37, 43f
    correcting errors, 44f
Ligatures, 37, *See also* Bindings; Duct tape evidence
Lips of decedent, 60f
Loaded weapons, 119, 120f
Locard's principle of exchange, 5

M

Masks, analyst equipment, 10
Medical examiner responsibilities, 2
Medical examiner's investigators, 157
Medical interventions, documenting, 9, 26f

Index

N

Neck of decedent, 53, 56–58*f*, 137*f*
Ninety degree blood drops, 73, 74*f*, 75*f*, 76*f*

O

Orange safety goggles, 18, 19*f*, 118*f*

P

Packaging
    gauze sleeves, 35*f*
    labeling, 32*f*, 34*f*, 35*f*, 37, 43*f*, 44*f*
    paper bags for large items, 38, 46*f*, 47*f*
    paper envelopes, 21*f*, 37
    plastic pouches, 45*f*
    swab boxes, 18, 23*f*, 33*f*, 37
    tamper-evident tape, 21*f*
Pants pockets, 38, 47*f*, 48*f*, 85, 89–91*f*, 140*f*, 146*f*, 150*f*, 152*f*, 154*f*
Paper bags, 38, 46*f*, 47*f*
Paper envelopes, 21*f*, 37
    labeling, 43*f*
Passive drops, 73
Penis of decedent, 53, 67*f*
Perpendicular blood drops, 73, 74*f*, 75*f*, 76*f*
Personal protective equipment, 9, 10
Picking, 10, 29–32*f*
Plastic pouches, 45*f*
Pockets, 38, 47*f*, 48*f*, 85, 89–91*f*, 140*f*, 146*f*, 150*f*, 152*f*, 154*f*
Preservation, 37–38, *See also* Packaging
    of clothing, 48–50*f*
    of duct tape, 41–42*f*
    of hands, 47*f*, 49–50*f*
    of knives, 38–40*f*
    of pockets, 47–48*f*
Proficiency testing, 5, 10*f*
Property crimes, 1

R

Rings, 68*f*
Rope bindings, 85, 109*f*
Rubber boots, 119
Rugs, bodies rolled up in, 117*f*, 155*f*

S

Safety goggles, 18, 19*f*, 118*f*
Safety priorities, 119
Saliva, 13, 18, 19*f*, 53, 118*f*
Scene screens, 18, 19, 25*f*
Semen, 18, 19*f*, 53, 118*f*
Sexual assault examination, 53
Shoe covers, 10
Shoelaces, 100*f*
Shoeprints, 50*f*
Shoes, blood patterns on, 75*f*, 76*f*
Shoulder swabs, 147*f*
Skin slippage, 11*f*
Smears, 79*f*, 80*f*
Socks, 63*f*
Standard operating procedures, 157
Strangulation-related evidence, 56*f*, 97*f*, 137*f*
Success rate of trace evidence analysis, 127, *See also* Trace evidence analysis, successful DNA recovery
Suitcases, bodies in, 115–116*f*
Supplies and materials, 7–9, 16–25*f*
    TECT rolling cart, 12*f*, 13*f*, 18–19
Swabbing, 13–15, 19, 53, 54–72*f*
    ankles, 62–63*f*
    arms and wrists, 64*f*
    body jewelry, 53, 65*f*
    breasts, 53, 61*f*, 149*f*
    bruises, 53, 69–70*f*, 154*f*
    cord bindings, 103*f*
    dismembered bodies, 71*f*
    face, 59*f*, 70*f*
    gloves, 88*f*
    knuckles or fingertips, 54–55*f*, 148*f*
    lips, 60*f*
    moistening swabs, 22*f*, 23*f*
    neck, 56–58*f*, 137*f*
    nonporous objects, 33*f*
    penis of decedent, 67*f*
    rings, 68*f*
    successful trace analyses, 137*f*
    turned-out pockets, *See* Pockets
    unusual stains, 66*f*, 153*f*
    using gauze, 18, 19, 34*f*, 35*f*, 109*f*, 145*f*
    voids in bloodstained areas, 53, 71*f*, 72*f*
Swab boxes, 18, 21*f*, 23*f*, 33*f*
    labeling, 34*f*, 37
Sweatbands, 17*f*, 18, 19
Swipe patterns, 73, 78*f*, 143*f*

T

Tamper-evident evidence tape, 21*f*
Tape lifts, 1, 10, 19, 24*f*, 27*f*, 28*f*
    clothing, 94*f*, 130–131*f*
    entire body, 95*f*
    labeling, 32*f*, 37
    preserving/storing, 13
    safety considerations, 122*f*
TECT submission form, 14*f*
TECT worksheet, 15*f*

Tents, 18
Tight areas, 119, 121*f*, 123*f*, 126*f*
Touch DNA, 1, 127
    body swabbing, 53
    duct tape evidence, 41*f*
    murdered officer's clothing, 86*f*
    property crime evidence, 1
    tape lifts, 28*f*, 130*f*
Trace DNA evidence collection team (TECT)
    administration, 157–158
    analyst training, 5–6
    call-out criteria, 2, 157–158
    Harris County program, 1–2
    scene interactions, 6–7
    supplies and materials, 7–9, 12*f*
    TECT rolling cart, 12*f*, 13*f*
Trace evidence analysis, successful DNA recovery, 127
    ankles, 141*f*
    bindings, 132–135*f*, 142*f*, 145*f*
    blood swipes on body, 143–144*f*
    breast swab, 149*f*
    bruise swab, 154*f*
    duct tape, 138–139*f*
    hair, 131*f*
    knife, 129*f*
    knot, 136*f*
    knuckle swabs, 148*f*
    murdered officer's clothing, 151*f*
    neck swab, 137*f*
    rolled-up rug, 155–156*f*
    shoulder swabs, 147*f*
    touch DNA on clothing, 130*f*
    turned-out pockets, 85, 140*f*, 146*f*, 150*f*, 152*f*, 154*f*
    unusual stains, 153*f*

Training, 5–6, 157
Tweezers, 10, 29–32*f*
Tyvek suit, 10, 16*f*, 119

U

Unique or unusual stains, 53, 66*f*, 153*f*

V

Void patterns, 53, 71*f*, 72*f*, 73, 86*f*

W

Waistbands, 53, 87*f*
Weapons, loaded, 119, 120*f*
Wet bodies, evidence collection techniques, 10, 29–30*f*
Wet items, 38, 46*f*
Wet scenes, 119, 124*f*
Wipe patterns, 73
Wrappings, 2, 85, 110–117*f*
    knots in, 113–115*f*, 136*f*
    processing, 110–112*f*
    rolled-up rugs, 117*f*, 155–156*f*
    successful trace analyses, 136*f*
Wrist swabbing, 64*f*

X

Xylene, 10

Z

Zip ties, 85, 106–107*f*, 142*f*

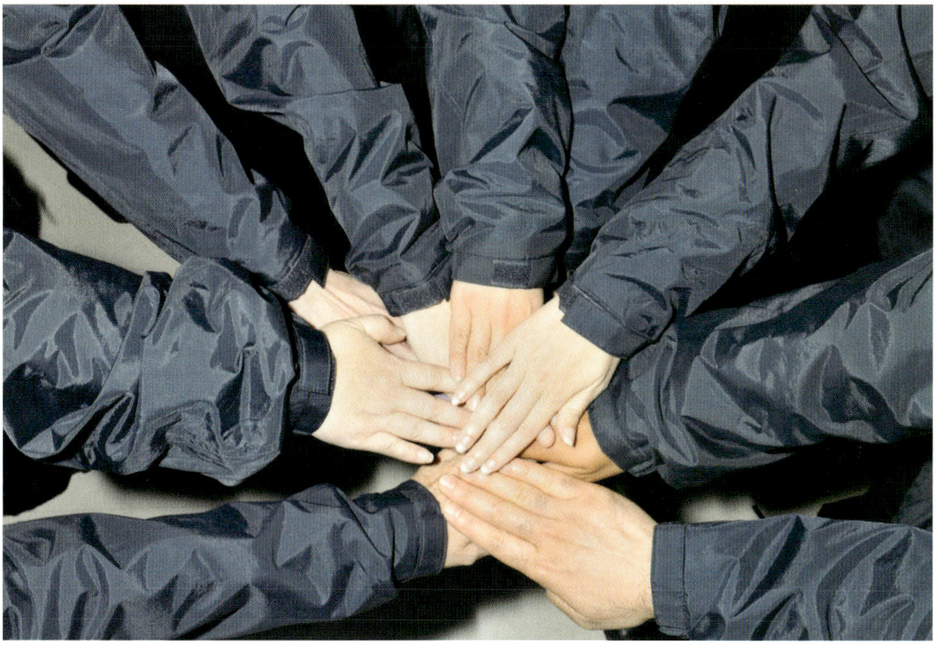